FARM TRACTORS

FARM TRACTORS

FEATURES MODELS FROM THE WORLD'S LEADING MANUFACTURERS INCLUDING
JOHN DEERE • IH • FORD • CASE • MERCEDES-BENZ • MASSEY-FERGUSON

MICHAEL WILLIAMS

THE LYONS PRESS

Guilford, Connecticut

An imprint of The Globe Pequot Press

The Lyons Press is an imprint of The Globe Pequot Press.

Originally published in 2002 by Amber Books, London

Project editor: Conor Kilgallon
Designer: Zoë Mellors
Picture research: Lisa Wren

Printed in Italy

2 4 6 8 10 9 7 5 3 1

ISBN 1-58574-503-0

The Library of Congress Cataloging-in-Publication Data is available on file.

CONTENTS

START OF THE POWER FARMING REVOLUTION

Tractor power has revolutionized farming methods. When the first tractors trundled off on threshing tours in the American Mid-West in the early 1890s, however, they were crude and unreliable. There was little evidence they would ever offer serious competition to the steam engine. Steam reigned supreme for another 20 years or so before tractors took the lead in the power farming revolution.

In the second half of the nineteenth century, it was the steam engine that provided farmers with their first alternative to muscle power for jobs ranging from ploughing to threshing. Until then, muscle power was provided by people who worked on the land and by large numbers of draught animals, including horses, oxen, mules and, occasionally, donkeys. For centuries, animals were used to pull the heavy loads, to plough and cultivate the soil, to power machines that threshed the grain and prepared feed for livestock and, in some cases, to give their owners a ride to and from the fields each day.

For the working animals, it was a life of toil, and some were literally worked to death. No doubt there were those that were well cared for and treated with at least a degree of sensitivity, but at a time when so many people experienced brutality in their own lives, there was probably little compassion to spare for the animals with which they

Like most tractors, Twin City machines, built by the Minneapolis Steel and Machinery Co., had become lighter and more versatile machines by the 1920s. The 1926 Twin City 21-32 tractor in the photograph was a four-cylinder model producing almost 36HP.

worked. As for the humans, although animals provided most of the power needed to grow and harvest crops in areas such as Europe and North America, there was still a huge demand worldwide for manpower in agriculture. Millions of people spent their working lives on farms doing jobs that were often strenuous and repetitive, and sometimes dangerous as well.

This was the way farming was organized for thousands of years. Viewed from the perspective of highly mechanized agriculture in the twenty-first century, it may have a rustic charm, but it was also an extremely inefficient way to produce food. The large numbers of working animals consumed significant quantities of the food they helped to grow, and the productivity per farm worker was so low that just a few centuries ago well over 50 per cent of the working population had to be employed on the land to provide enough food to meet the needs of a country such as Britain or France.

On British farms, the daily work rate for a skilled ploughman and two or three horses was an acre (0.4 hectares) per day in medium to heavy soils, increasing to 1.5 acres (0.6 hectares) daily on light, easily worked land. The average horse-power of new tractors sold in Britain during the early 2000s was almost 120HP, and a ploughman with this size of tractor would expect to plough two to three acres (0.8 to 1.2 hectares) per hour.

Thanks to the development of the steam engine and the tractor, few people

in the developed world experience the stresses and strains of farming solely with muscle power. It remains, however, the daily reality for millions of people in developing countries where food production is still limited to the pace of a team of oxen or the physical strength of a farmer and his family.

Power farming, based first on the steam engine and then on the tractor, has achieved a massive increase in farming efficiency and productivity. With mechanized agriculture, just two or three per cent of the working population produce enough to feed the other 97 per cent or so, with draught animals making occasional nostalgic appearances at old-time farming events and traditional ploughing matches.

THE START OF THE REVOLUTION

Britain had already established a clear lead in the use of steam power in factories and in the mining industry, and the first farms to use steam power were also British. The power farming revolution started in a modest way in 1798, when John Wilkinson, a wealthy businessman, installed a stationary engine on his farm near Wrexham in north Wales. This is the first recorded example of steam power being used on a farm, and Wilkinson used it to power a threshing machine, probably replacing one or two horses that would previously have been used for the same job.

Steam engines were inefficient and extremely expensive in the late eighteenth century, and it is unlikely that Wilkinson

Steam engines, together with many millions of hard-working horses, oxen and mules, provided most of the farming industry's power needs before the development of the tractor. This more efficient power source helped to produce food more abundantly and more cheaply than ever before.

would have expected to cover the cost of the engine through just a few weeks of threshing work each year. He did, however, have close business links with one of the leading steam engine manu-facturers; using a steam engine on his own farm may have been an attempt to encourage other landowners to follow his example and invest in steam.

The next reference to a steam engine on a farm comes from Scotland in the following year when 'a respectable farmer' in East Lothian was using a stationary engine for threshing. In this case, the farm was close to a coal mine, from which the farmer was able to collect his fuel from the pit-head at a reduced price. There are more reports of engine installations on farms in Britain during the next 40 years, but the numbers were small and the majority of farmers remained unconvinced of their usefulness. In some parts of the United States, farmers were apparently more willing to switch to steam power. A survey of agricultural steam engines in 1838 showed Pennsylvania and Louisiana leading the trend with steam power, with 274 engines already installed on Louisiana farms and estates, where they powered cane-crushing equipment for sugar production. Some of the engines were imported from Britain, but American manufacturers were expanding rapidly and would soon dominate the US domestic market.

In his book *Early Stationary Steam Engines in America*, Carroll W. Pursell quotes negotiations in 1812 between a Louisiana sugar estate owner and Benjamin Latrobe, one of the first of the American steam engine manufacturers. The engine was needed to power a sugar mill, and the price Latrobe quoted for an engine with a 30cm (12in) diameter cylinder was $2500. This figure would leave Latrobe with no profit at all, he claimed, but it would cover some of his overheads, and it might lead to further orders on which he could make a profit – which suggests that Mr Latrobe was a persuasive salesman. Most of the steam engines installed on Louisiana farms replaced the animal

power previously used for the crushers, and Pursell quotes another survey in 1840, when there were an estimated 400 steam-powered mills in the state, leaving 354 powered by animals.

FROM STATIONARY TO PORTABLE

In spite of the limited acceptance of stationary steam power on British farms and more impressive statistics from the sugar estates of Louisiana, the commercial impact of the agricultural steam engine was still minuscule even by the late 1830s. This was 40 years after John Wilkinson first showed how steam power could be used for threshing. Nevertheless, the technical breakthrough that made steam power available on many thousands of farms was on its way.

The breakthrough came when a steam engine was mounted on a chassis and four wheels so that horses could

tow it from farm to farm. This was the portable engine, and it took a surprisingly long time for such a logical idea to be developed. The portable engine made a significant difference to the economics of using steam power for farm work, as it could be operated by contractors, meaning that groups of farmers in the same area could share the cost and use of an engine.

Several British companies were developing portable engines at about the same time, but credit for being the first to demonstrate the idea usually goes to J. R. & A. Ransome of Ipswich, now a

Most of the steam engines used on farms were portables, used mainly for stationary work, such as threshing, and were pulled from farm to farm by teams of horses. This picture clearly shows the metal seat for the driver and the wooden pole where the horses were hitched.

subsidiary of the Textron company and known as Ransomes. It took a portable engine weighing 1.75 tonnes (1.72 tons) to the 1841 Royal Show, where it was used in a threshing demonstration and described as the 'great novelty' of the show. The Ransomes portable was said to produce as much power as five horses. A special design feature was a pipe taking the waste steam into the chimney, where it mingled with the smoke from the fire to extinguish sparks that might be released to cause a fire in the heaps of threshed straw.

In the following year, Ransome followed its success with the portable by building the world's first self-propelled agricultural steam engine, which it took to the 1842 Royal Show. The self-propelled engine, the immediate ancestor of the steam traction engine, made a slow commercial start, but the portable was an immediate success. By 1851, just 10 years after the first demonstration, the number of manufacturers of portable engines in Britain had reached at least a dozen and there were an estimated 8000 portables on UK farms.

Portable engines were also attracting interest in the United States, where the potential market for agricultural steam power was much bigger than in Britain. It is likely that some engines were built on a one-off basis before commercial production started, but the surviving records suggest that at least two manufacturers were offering portable engines for sale by 1849. Charles Hoad and Gilbert Bradford of Watertown, New York, were awarded a medal when one of their engines was shown at the 1851 New York Fair. A. L. Archambault's 'farm engine' was built in Philadelphia and was offered in three sizes with power outputs ranging up to 30HP, according to some reports.

A GROWING MANUFACTURING BASE

More US and Canadian manufacturers flooded into the market from the 1850s onwards, building both portable and traction engines. The first J. I. Case engine was built in 1869, when the demand for steam power was beginning to expand rapidly, and Case became the United States' largest manufacturer of portable and traction engines and also the biggest worldwide. The table below shows how Case production figures reflect trends in the market, including

Cable ploughing and cultivating were developed to overcome soil compaction and other problems caused by heavy traction engines. Two cable ploughing engines with a winding drum were parked at opposite sides of a field, and an attached plough or cultivator pulled to and fro between them.

the rapid collapse in sales when tractor power took over. Case built its last steam engine in 1924, when the production total had reached more than 36,000 engines in 55 years.

As well as increasing production, the steam engine manufacturers were also

CASE PRODUCTION

J. I. Case Annual Production Figures for Portable and Traction Engines

Year	Number produced
1882	506
1892	572
1902	1574
1912	2252
1922	153

J. I. Case Threshing Machine Co. was one of the world's biggest manufacturers of agricultural steam engines in the early 1900s. By 1924, the company had become one of the few traditional steam engine manufacturers to make the switch to tractor production.

making design improvements to boost performance and efficiency, and an indication of the progress in efficiency is available in the results of trials carried out annually by the Royal Agricultural Society of England (RASE). The trials measured fuel efficiency or the weight of coal used per horsepower-hour of power output, which means the fuel burnt to produce one horsepower for a period of one hour. They were organized with great care in order to ensure the results were as accurate as possible and to allow results for different years to be compared. The fuel for the engines was supplied from the same coal mine each year, and an analysis of the coal was published to make the results as useful as possible.

The results for the winning portables in six years of trials between 1849 and 1855 are shown in the table below, which was originally published in *The Engineer* in 1856.

A FARMING SOLUTION?

By the 1850s, steam power on farms was still restricted to stationary work, including threshing and driving machines for crushing or milling grain

for feeding livestock. This was the same type of work that steam engines had first handled 50 years previously. Steam was already providing power for manufacturing industries and for mining, and steam-powered railways and ships were revolutionizing transport; however, agriculture – the biggest industry of them all – still relied on horses and oxen for many important jobs, including ploughing, cultivating and harvesting.

It was a challenge that attracted both engineers and farmers, who invested large amounts of time and money seeking ways to make steam power

The first Case tractor was an experimental model built in 1892. Based on the chassis and wheels of a Case traction engine, its power unit was a twin-cylinder petrol engine. Reliability problems persuaded Case to abandon the tractor experiments and concentrate on its steam engines.

ENGINE EFFICIENCY

RASE Fuel Efficiency Trial Results

Year	Performance of winning engine (coal burnt per horsepower-hour)
1849	11.50 lb (5.22 kg)
1850	7.56 lb (3.43 kg)
1852	4.66 lb (2.11 kg)
1853	4.32 lb (1.96 kg)
1854	4.55 lb (2.07 kg)
1855	3.70 lb (1.68 kg)

available for field work. Most of the development work was in the 1850s, and many of the experiments in the United States, Britain and France produced expensive failures. The breakthrough came when John Fowler and others in Britain developed the cable ploughing system, using a steam-powered windlass to pull a plough attached to a cable to and fro across a field. Cable ploughing was popular in Europe, but failed to attract interest in the United States and Canada, where soil conditions in some areas allowed steam engines to plough by direct traction with the plough hitched to the rear of a traction engine. As the end of the nineteenth century approached, the

This is a replica of the tractor built by John Froelich in 1892. It was based on surviving photographs and engineering drawings of the original and was built to feature in a John Deere historical film. Most details are reasonably accurate, but it has a much later John Deere twin-cylinder power unit.

future of the big companies specializing in agricultural steam engine production must have looked secure, and it is unlikely that the arrival of the first tractors would have caused much anxiety.

EVOLUTION OF THE TRACTOR

While it was in Britain that the first steam engines were developed, the evolution of the tractor began in the United States and later spread to Europe. Credit for building the first tractor is usually given to John Charter of the Charter Gasoline Engine Co. based at Sterling, Illinois. In 1889, Charter mounted a big, single-cylinder petrol or gasoline engine made by his company on the wheels of a Rumely traction engine.

The tractor was taken to farms near Madison, South Dakota, where it was used to drive a pulley belt powering a threshing machine. The performance of the tractor must have been satisfactory because Charter's company received orders to supply a further five or six

tractors to farmers or contractors in the same area.

Competition for the Charter arrived in 1892 when at least three more experimental or pre-production tractors came onto the scene, all designed for threshing work and all built on the running gear of steam traction engines, with a slow-revving petrol engine to

FROELICH

Manufacturer: John Froelich
Model: n/a
Production started: 1892
Power unit: Van Duzen single-cylinder petrol engine with 35.3-litre or 2155 cu. in capacity
Power output: 16HP
Transmission: Exposed gear drive with two forward gears and one reverse

provide the power. Traction engine wheels and drive gears provided a readily available base for the engine, and it was a logical starting point for the early tractor pioneers.

One of the 1892 arrivals was the Capital tractor made by the Dissinger brothers from Wrightsville, Pennsylvania. They used an engine built under licence from the Otto company in Germany to power their tractor, which was designed for threshing. Little more was heard of the brothers' first tractor venture, but the Dissinger family returned to the tractor market a few years later with a new Capital tractor which proved popular in the early 1900s. A more significant name in the list of tractor pioneers in 1892 was the J. I. Case Threshing Machine Co. It mounted a twin-cylinder, four-stroke Patterson petrol engine with 20HP rated output on a set of traction engine wheels and axles – made presumably by Case – and this was used as a test vehicle.

A NEW COMMITMENT

The fact that the leading agricultural steam engine company took an interest at such an early stage in a potentially competing power source shows impressive foresight, and the fact that it also

quickly abandoned the project was almost certainly the correct commercial decision. In the 1890s, steam engines were already benefiting from well over a century of technical and commercial development, and they had established a reputation for reliability and a reasonable level of efficiency. Petrol engines were still at a very early stage of development and were far from reliable, with primitive fuel and ignition systems that were notoriously temperamental. The Case engineers soon discovered that the petrol engine on their tractor lacked the reliability of their steam engines, and it was this that brought the tractor development programme to an abrupt end.

Case showed equally good judgement when it decided to start a new development programme for tractors almost 20 years later. By about 1910, when its new tractor project began, production of Case traction and portable engines had reached record levels; however, the company decided this was the right time to move into the tractor market and, as usual, its timing was perfect. While many rival steam engine manufacturers simply continued building traditional engines for what soon became a shrinking market, Case was

It was the Waterloo Boy models R and N that brought John Deere into large-scale tractor production, although they were never sold under the John Deere name. The Waterloo Boy, with its traditional steel frame, remained in production until John Deere had designed a new model.

ready to move into the tractor market just in time to catch the massive sales boom encouraged by the 1914–18 war. It was also able to take advantage of almost 20 years of technical improvements in petrol engine design and subsequently build tractors that were much more reliable than anything it could have offered in 1892.

FROELICH ENTERS THE FRAY

The third member of the group of tractor pioneers in 1892 was John Froelich, who lived in Froelich, Iowa, a small town named after his parents. At an early stage in his career, Froelich was running a steam-powered grain elevator plus a feed mill, but he also built up a business as a contractor, operating his own well-drilling equipment and also working with a threshing crew on farms in the Dakotas. In 1890, he bought a petrol engine from the Charter Gasoline

Engine Co. It produced 4.5HP rated output and had one horizontal cylinder.

Froelich bought the engine to power the drill he used for his well-boring business, and it may have encouraged his idea to use a similar engine in a tractor. It is also possible that he may have heard about the Charter tractors when he was working on farms in the Dakotas. He decided to build his own tractor, and the power unit he chose was a single-cylinder Van Duzen petrol engine made in Cincinnati, Ohio. While most of the early petrol engines at that time had a horizontal cylinder, the Van Duzen engine was a vertical design.

The cylinder of the Van Duzen engine was massive, with the 356mm (14in) bore and stroke providing 35.3 litres (2155 cu. in) capacity and producing a decidedly modest 16HP output. The biggest engine currently available in a John Deere tractor is a six-cylinder diesel with a turbocharger and intercooler, delivering well over 400HP from 14 litres (854 cu. in).

Froelich and William Mann, one of his employees, made a wooden chassis to carry the engine, and this was mounted on traction engine running gear. Drive shafts, gear wheels and other components were bought from the Robinson steam engine company in Richmond, Indiana. The tractor was completed in 1892 and, after a number of modifications and adjustments, worked successfully.

John Froelich designed his tractor with the engine in the middle and a platform at the front for the driver. This gave an almost unobstructed forward view, which was a big improvement on the usual steam traction engine layout

with the driver at the rear. The operator had to stand while driving the tractor, partly because the steering wheel could be operated only from a standing position and partly because no seat was provided. The only concession to driver comfort was a wooden container below the steering wheel to hold a large can of drinking water.

The main fuel tank was at the rear of the driver's platform, and there was a lever-operated pump to transfer petrol from the main tank to a small cylindrical tank high above the engine. The engine was not equipped with a fuel pump, and fuel was supplied by gravity feed from the overhead tank. The transmission was a gear drive from the engine to a large-diameter ring gear on each of the rear wheels. All the gear wheels were completely exposed to dirt and dust, but on the later production version most of the transmission was enclosed.

A NEW TRACTOR COMPANY IS BORN

Froelich bought a new Case threshing machine and took it and the tractor to South Dakota. The records he kept of this threshing tour show the equipment was working for 52.5 days, and during that time it threshed 62,000 bushels of wheat and other small grains. No major breakdowns were reported, and, after Froelich returned from Dakota in November, enthusiastic reports of the tractor's performance attracted the interest of a group of businessmen in Waterloo, Iowa. Froelich was invited to a meeting where it was agreed to form a new company in Waterloo to build tractors based on his design.

The company was established in January 1993 and was called the

The Waterloo Boy badge as it appeared on the models R and N tractors is one of the most famous in tractor history, with its smiling all-American Iowa farm boy. A new identity based on the Overtime name was adopted when some of the tractors were shipped to Britain.

WATERLOO BOY

A sales leaflet for the Waterloo company's new tractors made it clear they were competing for sales against 'the old-fashioned, cumbersome and complicated steam traction engine'. The leaflet included a list of 19 advantages over its rival, and these were reproduced in volume II of the Two-Cylinder Collector Series published by the Two-Cylinder Club. The claimed advantages were:

1. No possibility of explosion.
2. No possible danger of fire.
3. No tank man and team necessary.
4. A high-priced engineer is unnecessary.
5. No early firing to get up steam.
6. No leaky flues.
7. No boiler repairs of any kind.
8. No boiler cleaning and breaking of handhole bolts.
9. No broken bridges on account of weight.
10. No waiting for steam.
11. No waiting for water.
12. No running into holes or other obstructions, because the operator stands in front and has full view of the road before him.
13. No time lost making long moves to take on fuel and water.
14. No time lost in turning the engine after the separator is uncoupled and left between the stacks.
15. No consumption of fuel before starting or after stopping.
16. No exact lining with separator necessary.
17. No runaway teams on account of 'steam blowing off'.
18. No long belt to contend with.
19. No stopping of the engine when changing from separator to traction.

WATERLOO BOY

positioning the tractor correctly to power the belt drive to a thresher.

Another reason why John Froelich's tractor is so important is that it was the original forerunner of the modern John Deere tractor range. Waterloo expanded to become one of the leading stationary engine manufacturers, but they later returned to tractor production with the Waterloo Boy range, available in various versions from 1912. When Deere and Co. decided to buy its way into the tractor market in 1918, the manufacturer it chose to take over at a cost of £2,350,000 ($3,760,000) was the Waterloo Gasoline Engine Co.

Waterloo Gasoline Traction Engine Company. It was based in Waterloo, with Froelich as the president, and, rather unusually, the Van Duzen engine from his first tractor was removed and used to power the new factory. Four new tractors were built, all based on his original design, but with a number of improvements, including steel instead of wood for the main frame.

Two of the Waterloo tractors were sold, but both were returned by their new owners because of mechanical problems. This was a serious setback for the company, and it decided to concentrate on making stationary engines instead of tractors. The engines, designed initially by Froelich, were a success; in 1895, the company name was changed to the Waterloo Gasoline Engine Co., omitting the word 'Traction'. At about the same time, Froelich left the company to begin a new career elsewhere.

This was not an auspicious start; however, in spite of the setbacks and John Froelich's departure from the company, his first tractor was a highly significant development. It was probably the first tractor to be equipped with reverse as well as forward gears, and this must have been a big advantage when

THE BEGINNINGS OF JOHN DEERE

The Waterloo Boy tractors that took Deere into the market were the R and N models. The Model R was introduced in 1915, with production continuing until 1919, and the Model N was available from 1917 until 1924. This means both Waterloo Boy models were the first

The final drive on the Model R was based on ring gears that were fully exposed to dust and mud, and wear must have been a problem on abrasive or stony soils. The ring gears on the Model R were smaller than those on the Model N, being little more than half the tractor's rear-wheel diameter.

tractors to be sold by Deere, even though they never carried the John Deere name. They also introduced Deere to the twin-cylinder horizontal engine layout that remained a successful feature of almost every John Deere production tractor for more than 40 years.

Although at first glance the R and N models look similar, sharing as they do the same layout of engine, transmission and cooling system mounted as individual units on a steel girder frame, there were important differences. Design changes introduced on the Model N included a two-speed transmission instead of the single-speed version found on the Model R. Also, while the big ring gears on the driving wheels of the Model N are almost the same diameter as the wheel itself, on the Model R version, these are little more than half the diameter of the wheel. The Model N radiator is mounted on the left-hand side of the frame, viewed from the driver's seat, but on most Model Rs it is on the right-hand side. The steering system is a less reliable indicator, as chain-link steering was fitted to all Model R tractors and to Model Ns built

before about 1920, but this was replaced by more accurate worm and sector steering from 1920 onwards.

As well as being the first John Deere tractors, the Waterloo Boys possessed other claims to fame. One of the distributors for tractors exported to Britain, where they were sold under the Overtime brand name, was Harry Ferguson – the Overtime almost certainly triggered the early development of the Ferguson System of implement attachment and control (see 'Designed for Performance'). In 1920, the Model N Waterloo Boy also became the first tractor to complete a Nebraska test. The results confirmed the 12-25HP rating at a sedate 750rpm.

HUBER ENTERS THE MARKET

The Van Duzen company, manufacturer of the engine for John Froelich's first tractor, designed and patented a tractor of its own in 1898. It was, of course, powered by one of its own single-cylinder vertical petrol engines and, like the Froelich tractor, it featured a gravity feed to deliver petrol from a holding tank high above the engine. Later in the

The Waterloo Boy tractors were not alone in retaining a steel-frame structure well into the 1920s. The Huber Manufacturing Co. in Marion, Ohio, started building its Light Four tractor in 1917. The Light Four proved popular with customers and was still available in 1928.

same year, the Huber Manufacturing Co. based at Marion, Ohio, made a successful takeover bid for Van Duzen and its engine and tractor interests. This brought Huber into the tractor market for the first time, and it is reported to have built a batch of 30 tractors based on the Van Duzen design. If this figure is correct, it probably makes Huber the world's biggest tractor manufacturer in the period before 1900.

Huber was already a leading manufacturer of farm implements and steam traction engines, and, having completed their initial production batch of tractors, they pulled out of the market for more than 10 years to concentrate on other products. They began making tractors again in about 1911, starting with the Farmer's Tractor, powered by a two-cylinder horizontally opposed engine

mounted in the middle of the chassis. The cooling tower for the engine was at the rear, the driver was positioned at the front and the tractor was mounted on a set of Huber steam engine wheels. Production continued with mainly medium-powered tractors using engines supplied by Buda, Stearns and Waukesha. Huber tractors remained popular throughout the 1920s and early 1930s, but production ended during World War II.

The Sterling tractor, which made a brief appearance between 1893 and

This was the third tractor built by Charles Hart and Charles Parr. It was tested in 1903 and was given an 18–30HP rating, with the power delivered from a two-cylinder engine. The tractor, with Hart-Parr's oil cooling system, is displayed at the Smithsonian Institution, Washington DC.

about 1895, was made at Sterling, Kansas, to the design of a Mr Hocket. The front view of the Sterling, with its large chimney-type exhaust, looked similar to a steam traction engine, and it has been suggested that this was a deliberate attempt to make it appear more familiar and less frightening to horses passing it on the road. The Sterling specification even included a whistle, useful for signalling to members of a threshing crew.

With the possible exception of the Huber company, tractor production remained a small-scale business until the end of the nineteenth century; however, during the early 1900s, the numbers of tractors built by some of the leading manufacturers grew significantly. One example was the Hart-Parr company named after its founders, Charles Hart and Charles Parr, who began designing and making engines while both were engineering students at the University of Wisconsin in Madison.

THE HART-PARR COMPANY

Hart & Parr was established in Madison before the two men graduated in 1896; when they left the university, they were ready to start building stationary engines commercially. The company name was changed to Hart-Parr in 1897, and, as the business expanded, it became obvious that it needed more space to increase its production capacity. The company was moved to

Charles City, Iowa, where Charles Hart had spent his childhood, and it had built its first tractor by 1902.

For tractor number one, Hart-Parr used one of its own twin-cylinder engines with a 30HP power rating and the oil cooling system the partners had developed while studying at university. The oil-cooled engine remained a Hart-Parr speciality until well into the 1920s, and most of the tractors with this feature were equipped with a distinctive rectangular cooling tower at the front of the tractor. Benefits claimed for oil cooling included avoiding the risk of frost damage. The oil also allowed higher operating temperatures, thereby improving the combustion of low-grade fuels such as paraffin or kerosene.

During its first 20 years of tractor production, the Hart-Parr company concentrated on making tractors suitable for big prairie farms in the United States and Canada. While the first production tractors were designed mainly for delivering power through the belt pulley for threshing, improvements to the transmission allowed some models built after about 1904 to be used for heavy-duty haulage work and later for direct ploughing.

Hart-Parr became one of the leading companies in the heavy duty end of the tractor market, and its range was topped by the 60-100 model, which weighed 26.5 tonnes (26 tons) and had rated outputs of 100HP at the flywheel and

60HP at the drawbar. Lighter tractors were introduced during the 1920s, including Hart-Parr's smallest model, the lightweight 10-20 introduced in 1921, which was powered by a two-cylinder engine. After three years, the 10-20 was replaced by the 12-24 E, which was in turn replaced, in 1928, by an improved H version of the 12-24.

A move back up the power scale brought the 28-50 model, powered by two of the 12-24 two-cylinder engines placed side-by-side to provide four-cylinder power. In 1929, Hart-Parr became part of the Oliver Farm Equipment Co., and this group was later taken over by the White Motor Co.

One of the companies providing competition for Hart-Parr at the heavyweight end of the market was the Advance-Rumely Thresher Co. of

---

STOPPING POWER

The priority in the early years of tractor development was reliability, and easy starting was much more important than stopping power.

Brakes were a rarity, and even the Model F Fordson started its production life in 1917 without brakes. At a time when driver safety was a low priority, lack of brakes does not appear to have concerned customers, operators or tractor manufacturers.

An exception was the first Ransomes tractor built in 1903 and featuring two braking systems. One was operated by a foot pedal that disengaged the clutch and then applied a brake to the transmission shaft, and the other was controlled by a hand lever to operate brakes on the rear wheels.

Later, when tractor speeds increased, transmission brakes became standard, operating on the rear wheels only. Since the 1980s some tractors have been equipped with brakes on all four wheels, and the pioneers of this trend include Same, Landini and the JCB high-speed tractor.

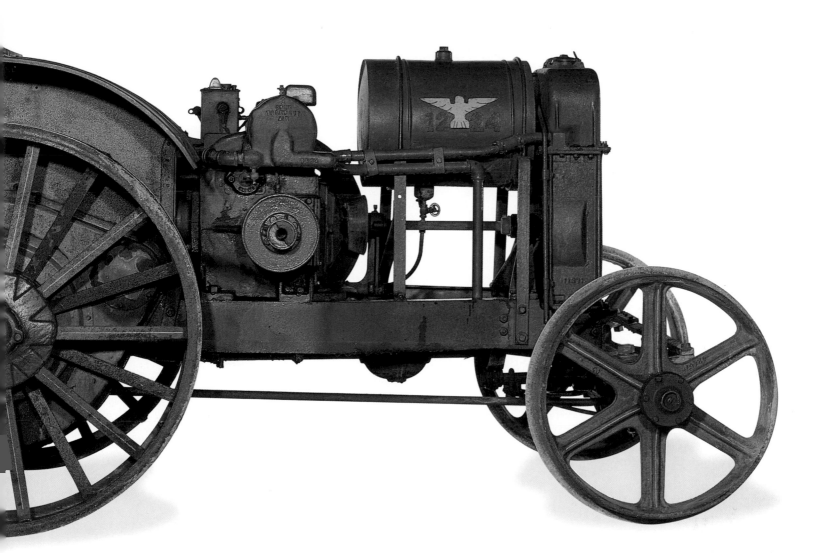

LaPorte, Indiana. It had already established a reputation in the farm equipment market for building steam engines in stationary, portable and traction versions, plus a highly successful range of threshing machines. Its first move into the tractor market came in 1908, when John Secor joined the company and developed the paraffin-burning engine from which the company's OilPull brand name came. (see 'Designed for Performance').

RUMELY MAKES ITS MARK

Like its Hart-Parr rivals, the new Rumely engines used oil instead of water for cooling, and most of the production tractors carried a distinctive rectangular cooling tower at front, which did little to improve forward visibility from the tractor seat. Rumely

tractors were popular, and one of their biggest successes was the Model E built for about 12 years from 1911 and equipped with a single-speed gearbox. The official power rating was 30-60, or 60HP on the pulley and 30HP at the drawbar, but these figures were easily exceeded when the Model E was tested at Nebraska, producing almost 50HP in the maximum drawbar pull test and just over 75HP on the belt pulley.

The 30-60 and the GasPull model with a 35-70 power rating were the top models in the Rumely range. The smaller versions included the Model H tractor introduced in 1919 and rated at 16-30HP. The output was achieved at 530rpm from a two-cylinder engine with 178mm (7in) bore and 216mm (8.5in) stroke. The Model H was one of the old-style models that disappeared in

Later Hart-Parr tractors acquired a more conventional layout and also some weight compared with the company's earlier designs. This is the H12-24 model displayed at the Farm Museum at Milton, Ontario. It was powered by a twin-cylinder engine with the horizontal layout favoured by John Deere.

1924 after the launch of a new range of smaller, lighter OilPulls featuring unitary construction for the first time in Rumely tractors.

Although the new models were given more modern styling, the distinctive rectangular cooling tower was retained – no doubt pleasing the many Rumely fans – but the new version was lower and less obtrusive. The first batch of four new models announced in 1924 started with the Model L, the smallest version, which had a 15-25 power rating. The

OilPull Model M had a 20-35 rating, and this was increased to 25-45 for the Model R. Although the last of the four new models, the Model S, was introduced as part of the lightweight range, by most standards it was still a heavyweight. It tipped the scales at almost 8.1 tonnes (8 tons) and had a 30HP rated output at the drawbar and 60 on the belt pulley. Nebraska test figures for the lightweight models show that Rumely was continuing to quote output figures conservatively; an example of this is the 25-45 Model R, which had maximum outputs of 32.6HP at the drawbar and more than 50HP on the pulley.

More new and updated models followed in the late 1920s, including the company's first and last six-cylinder model, the Rumely 6; however, these failed to halt a decline in sales. By 1931, when Rumely was facing serious financial problems, the company was taken over by Allis-Chalmers.

DEERING AND McCORMICK

While most of the emphasis in the earliest days of the American tractor industry was on replacing steam engines for threshing work, there were exceptions. These included Deering and McCormick, the companies that joined forces in 1902 to form the Chicago-based International Harvester Co.

Deering and McCormick both used small petrol engines to power tractors designed as self-propelled mowing machines, and they were both available

This front view of the H12-28 shows the steering wheel offset to the driver's right-hand side. This sales feature was favoured by many of the leading US manufacturers, and it was designed to give a better view from the driver's seat when working with a plough, helping to keep furrows straighter.

in small numbers from about 1897. The Deering mower was powered by a two-cylinder horizontal engine and was equipped with a 1.5m (5ft) wide cutter-bar. The McCormick version was called the Auto-Mower and used a 6HP petrol engine. Both were designed as small, lightweight machines based on a three-wheeled design, with the single wheel at the front. The Auto-Mower was equipped with tiller steering designed for one-handed operation, while the Deering mower used a small-diameter car-type steering wheel.

The Deering and McCormick companies were both actively developing export sales to Europe by the end of the nineteenth century; in 1901, their self-propelled mowers competed on level terms in an official test in a field near Paris. Both apparently performed well, with the McCormick Auto-Mower working at a steady 8km/h (5mph). A report by the judges also commented favourably on the fact that it was easy to remove the engine from the Deering machine and use it as a stationary power unit. This seems a somewhat odd comment, as it would have been more convenient to have been able to leave the engine mounted on the mower for stationary use.

One of the most ingenious but short-lived attempts to bring low-cost tractor power to smaller farms came from the Adams Husker Company of Marysville, Ohio, when it announced its Little Traction Gear model in 1909 or 1910. This was a tractor with no engine, which obviously helped to reduce the price to the farmer. It was supplied with

an empty space where the engine should be, and it was aimed at those farmers who already owned a suitable slow-speed stationary engine which they could mount in the space provided.

A chain and sprocket supplied with the tractor was used to link the engine to the tractor's transmission, supplying power to the pulley belt and by another chain drive to the rear wheels. C. H. Wendell in his *Encyclopedia of American Farm Tractors* says that the Little Traction Gear was available in three sizes, one designed for engines up to 9HP, a medium-sized version for engines of 13HP or less and the largest, which was built to accept up to 20HP.

THE BIRTH OF BRITISH TRACTORS

Britain was the first European country to experiment with tractor power, and the first tractor to be built commercially in Britain arrived in 1896. It was designed and built by Richard Hornsby and Sons of Grantham, Lincolnshire, and its full official name was the

Hornsby-Akroyd Patent Safety Oil Traction Engine. The makers promised four versions of the tractor powered by engines with 16, 20, 25 and 32HP output, but it is unlikely that all of these were built. The engine was a semi-diesel based on a Stuart and Binney design, built by Hornsby under a licence agreement. The layout was horizontal, and it was started by using a blow lamp and ran on paraffin with the power delivered through a transmission with three forward gears and one reverse.

Although the Hornsby tractor was designed like a traction engine for stationary work, it was also suitable for heavy haulage jobs on farms or on public roads, as indicated by the extremely strong chassis and three-speed gearbox. The sales leaflet emphasized the tractor's haulage capabilities, suggesting that the 16HP model would handle a 20.3- to 25.4-tonne (20- to 25-ton) load on level ground. The 32HP model was claimed to be suitable for loads weighing up to 50.8 tonnes (50 tons).

The cost of developing a brand-new engine can be high; when the Hart-Parr company decided to add a new 50HP model to its list of tractors in 1927, it found an ingenious way to avoid the outlay of a new engine. Instead of developing a new 50HP power unit, it mounted two of its existing 24HP engines side by side to produce the new 28-50 model.

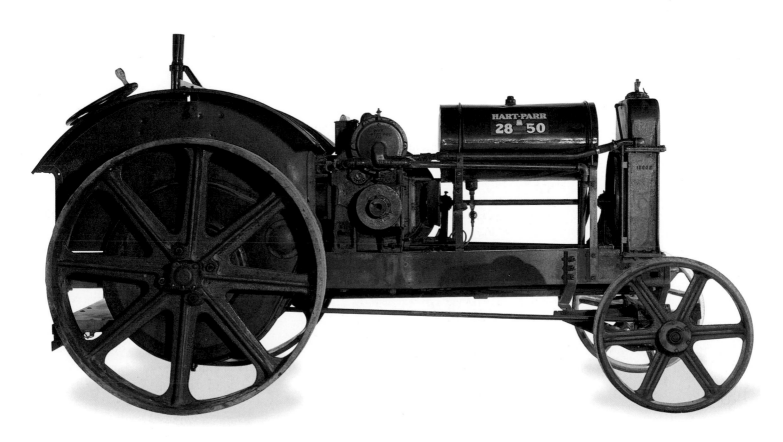

When the new tractor was demonstrated before the machinery awards judges at the 1897 Royal Show, they were favourably impressed by its manoeuvrability and its ability to cope with the test circuit that included driving over railway sleepers and crossing soft ground. They obviously compared the Hornsby against steam traction engines and included 'non-liability to explode' in their list of reasons for deciding to give the new tractor a silver medal. Steam engine comparisons were also prominent in a description of the tractor published in 1896 in a leading journal, *Implement and Machinery Review*.

'The driver has a good deal easier time of it than in the case of a steam-engine,' said the *Implement and Machinery Review*'s report. 'There is no fire to be frequently stoked, nor are there any water or steam gauges to be kept under supervision. Indeed, the duties are so comparatively light that one man can

Hart-Parr's big 28-50 had the same offset driving position featured on its H12-28. Production of the 28-50 started in 1927 and continued until about 1929, when the tractor range was completely updated after four companies merged in 1929 to form the Oliver Farm Equipment Co.

easily undertake the driving without any assistance, which, of course, means a considerable saving to the user.' Another feature praised in the report was the fact that the engine's exhaust system had been 'rendered silent' to avoid frightening horses.

A NEW COMMERCIAL MODEL

Prices for the Hornsby tractor started at £500 for the 16HP model, which was supplied with a pair of carriage lamps for driving on the road, a tool kit including spanners plus a hammer and chisels, a waterproof cover, and a rear-mounted winch and a cable for pulling timber. There was also a bucket, and this was used to top up the water in the cooling system, which evaporated or leaked about 250 litres (65 gallons) during a day's work.

Only one of the tractors was sold in Britain. The customer, who owned a large estate in Surrey, became the first person in Britain to buy a tractor. Another three or four of the tractors were exported to Australia, and one of these was recently brought back to its country of origin by a vintage tractor enthusiast. Richard Hornsby also entered one of its tractors in an evaluation test organized by the secretary of state for war. The British

Early versions of the Rumely OilPull tractors were noted for their large size, heavy weight and the distinctive cooling tower at the front. The tractor in the photograph is a Model H, introduced in 1919 with a 30HP twin-cylinder engine and the familiar, but much-reduced, cooling tower.

Army wanted a tractor to replace some of its steam traction engines to deal with transport work such as moving heavy guns and other equipment. There was a £1000 prize for the winning manufacturer and, more importantly, the possibility of a lucrative contract. The Hornsby tractor won the prize, but the contract was never awarded, and the company's subsequent efforts to interest the army in a track-laying version of the tractor – years before the army decided it urgently needed tanks – also failed to win an order.

The design of the Hornsby tractors was obviously influenced by steam traction engine ideas, but other British

manufacturers were designing their tractors for a wider range of jobs. This was the approach used by the Marshall company of Gainsborough, Lincolnshire, when it announced its first tractor in 1906. It weighed 4.6 tonnes (4.5 tons) and was designed mainly for export, achieving modest success in Canada against mainly US competition.

A paraffin-burning, twin-cylinder engine developing 30HP powered the early versions of the Marshall, and the engine was water-cooled, losing 9.1 to 13.6 litres (two to three gallons) each day through evaporation and what Marshall called 'inevitable leakages'. Hot water from the individually jacketed

engine cylinders was circulated to the top of the radiator, moving down to the bottom of the radiator as it cooled, and it was then piped to the main tank at the rear before making a return journey to the cylinders.

THE FIRST MARSHALL

The first Marshall tractor was equipped with a pulley belt for stationary work and with a three-speed gearbox for haulage. However, when the tractor was announced, the manufacturers gained useful publicity by demonstrating its performance in a non-stop ploughing marathon. The marathon lasted 24 hours and, during that time, the Marshall used

200 litres (44 gallons) of paraffin to plough nine hectares (22 acres).

There were also some British manufacturers offering a much more versatile approach to tractor design in the early 1900s, and many agricultural historians regard these as the real ancestors of the general-purpose farm tractor. All the tractors they built were small enough and light enough for field work in a wide range of soil conditions, and they were designed as a replacement for the farm horse, instead of competing with the steam engine.

Some members of this small group of pioneers made only a brief appearance in the tractor market, quickly deciding

A pre-heater on Rumely OilPull tractors, as
seen here on a Model H tractor built
between 1919 and 1924, used waste heat
from the engine exhaust to raise the
temperature of paraffin or kerosene fuel for
easier combustion. The use of cheaper,
lower-grade fuels was a major attraction.

to concentrate on other products, and this group includes Ransomes, the company which put the steam engine on wheels in the early 1840s. They built a prototype tractor in 1903, using a 20HP Sims engine designed for the car industry, and the general layout appears to have been influenced by early twentieth-century car design. A three-ratio gearbox controlled the speed of the belt pulley, as well as the forward speed, and this gave the driver the choice of 220, 450 or 1000rpm on the pulley with the engine set at its rated speed.

As well as having a pulley for stationary work, the Ransomes tractor was also designed for direct traction ploughing, with a claimed 0.2 hectares (0.5 acres) per hour work rate and 6.8 litres (1.5 gallons) per hour petrol consumption. The tractor was also said to be capable of pulling a '7- to 8-ton load' up a 'steep' hill at 11.3km/h (7mph).

Drake and Fletcher, an engineering company based in Maidstone, Kent, demonstrated its first tractor in 1903. It built its own three-cylinder petrol engine to power the tractor, claiming a 16HP output, and the tractor was designed to provide maximum versatility, as the following report in the 3 July 1903 issue of the *Hardware Trades Journal* suggests.

The tractor is 'designed for all kinds of farm and general estate work, including ploughing, cultivating, reaping, binding, mowing, hop washing, etc. and is also capable of being used for stationary work'. The reference to hop washing is a result of the fact that the tractor was built in Britain's leading hop-growing area.

NEW POSSIBILITES IN TRACTOR POWER

Professor John Scott, an agricultural college lecturer who turned to tractor development, was one of the first people to understand the possibilities offered by tractor power. In 1906, he told a farmers' club meeting in Scotland that the tractor could do all the work on a farm in about half the time required with horses and with about one-third of the manpower. 'The time seems fast approaching when motor power will be universally used on the farm, and farm labourers will know more about motors than about horses,' he told his sceptical audience, many of whom were probably making money by selling oats or hay for horse feed or by breeding horses for sale to other farmers.

Scott, who lived near Edinburgh, built a series of tractors with features that were years ahead of their time, and he

achieved little recognition or commercial success. His first tractor was displayed at the 1900 Royal Show and carried a cultivator/drill combination mounted on the rear. The cultivator, with sets of rotary tines powered by a chain drive from the tractor's rear axle, was probably the earliest ancestor of the power harrows that would eventually become popular 60 years later. Scott's idea of combining a seed drill and a cultivator to do two jobs in one operation was about 70 years ahead of its time. Another Scott tractor announced in 1904 featured a power take-off designed to drive a front-mounted mower or reaper. The power take-off did not become widely used for at least 30 years, and the idea of mounting equipment on the front of the tractor took another 70 years to achieve limited acceptance.

H. P. Saunderson of Elstow, Bedford, was another member of the group of highly inventive British pioneers who helped to develop the tractor as a general-purpose power unit, and in commercial terms he was easily the most successful. He used the name Universal for his tractors to stress their versatility, and his first success came in 1906 when an improved version of his first Universal model powered by a 30HP engine won a Royal Show silver medal.

A special feature of early Saunderson designs was a load-carrying platform over the rear section of the main frame, allowing the tractors to be used as load carriers as well as for pulling equipment and powering machines from the belt pulley. The load platform was designed to carry up to 2 tonnes (2 tons) and could be manually tipped, helping to make the Saunderson tractor an effective replacement for two or three horses. To prove its point, the company invited editors of leading farming magazines to a demonstration in 1906.

The event started after lunch with a Saunderson Universal pulling two 1.8m (6ft) binders to harvest 0.8 hectares (two acres) of wheat. When the crop was cut, the binder was unhitched and the same tractor pulled a threshing machine into the field. The tractor then worked with its load platform to carry the sheaves to the thresher. Once this job was finished, it was used as a stationary power source to drive the threshing machine. The pulley was then used again to drive a grinder to turn the freshly harvested grain into flour. At this stage, a baker took over, turning the flour into dough

and baking loaves of bread in an oven, while the tractor went on to plough the recently harvested area, cultivate it and then sow the seed for the next crop of wheat.

Within five hours, one crop had been harvested and prepared for baking, and the next year's crop had been sown, all with the power of one tractor, and the journalists were able to eat freshly baked bread from the newly harvested crop. Now, almost 100 years later, the fact that one tractor can do so many different jobs does not seem surprising; however, in 1906, it was an impressive

display of the way in which farm mechanization would soon develop.

Saunderson later abandoned the load platform – transport work was probably easier with the tractor pulling a trailer – and within a few years it had become Britain's biggest tractor manufacturer

The OilPull brand name was designed to advertise the Rumely tractors' ability to burn lower-grade fuel, a feature which helped to boost sales of the OilPull. This was further emphasized by the phrasing of the guarantee displayed on the side of the cooling tower of this Model H tractor.

with a flourishing export business. Its best-selling model was the 25HP Universal it built during World War I, but competition from cheaper, more up-to-date models eventually put it out of business; the Saunderson company was taken over by Crossley of Manchester, a leading engine manufacturer.

THE IMPACT OF DAN ALBONE

The most innovative of the early tractor pioneers in Britain was Dan Albone. His name might well have been remembered alongside those of Harry Ferguson and Henry Ford; however, though findly remembered by just a few tractor history enthusiasts, he is now almost forgotten.

The operator's platform on this Rumely shows how uncomfortable and potentially dangerous a tractor driver's job could be. On this tractor, as on a number of OilPulls, there is no seat for the operator. There is also nothing to stop the operator accidentally stepping or falling backwards off the narrow platform and into the path of the implement following the tractor.

Albone was raised on a small farm or market garden and, like Ferguson and Ford, his engineering skills were all self-taught. He started a business as a bicycle manufacturer and, in 1898, designed and built his own car. His interest in tractors started in about 1896, but he did not start building his first tractor until 1901. He chose a three-wheeled design, with a single wheel at the front, and the engine was mid-mounted and powered the single-speed gearbox through a cone clutch. A large metal tank beside the driver at the rear of the tractor held the water for the cooling system and also put plenty of weight over the driving wheels. It was powered by various car-type engines, most of them made by

Payne and Bates, and the power started at about 8HP and was increased steadily to provide more performance until it reached 24HP in 1913. The tractor was called the Ivel, which was the trade name Albone had used for his bicycles and also the name of the river running through Biggleswade, Bedfordshire, where his business was based.

Few of Britain's farmers were ready for tractor power in the early years of the twentieth century. The big acreage farms and estates were equipped with steam engines and the rest relied on horsepower, and British tractor makers were forced to find overseas markets for most of their sales. By 1906, Ivel tractors had been sold in 18 countries from Canada to Australia and from Nigeria to Cuba, and the successes included an order for 18 tractors for the Philippines.

The Ivel was described as 'undoubtedly the great attraction' at the 1904 Paris Show. Furthermore, a Canadian farmer was quoted as describing it as 'the new farmer's friend', and a feature about tractor development in America and Britain in a leading publication in Argentina said the Ivel was 'the most successful agricultural motor yet placed on the market'. After a Ministry of Agriculture demonstration in Italy, the organizers were so impressed by the Ivel that they gave Albone a special medal to commemorate the event.

As well as developing export markets, Albone showed extraordinary imagination in seeking new opportunities to sell his tractors in Britain. He kept an Ivel tractor and a range of machinery on land near the factory where demonstrations of farming by tractor power

Marshall was one of Britain's leading manufacturers of steam engines, and it made a determined effort to move into the tractor market. Because of its steam engine, it concentrated on big tractors, including a 30HP model announced in 1906 powered by a paraffin-burning engine.

were held once every fortnight. One of his tractors, complete with crew dressed in firemen's uniforms, was demonstrated as a fire engine, using the belt pulley to power the pump. He also turned an Ivel tractor into a military ambulance, with steel cladding to protect the driver and two rear doors made of steel plate opening outwards to provide some protection for stretcher bearers walking behind the tractor. Real bullets were fired at the tractor to test the armour plating during a demonstration for the army, and Dan Albone showed how the pulley could be used to power ice-making plant or equipment for purifying water for a field hospital. He also demonstrated the tractor's ability to haul medical supplies over rough

ground – despite this, the military chiefs decided they still preferred horses.

Albone died in 1906 at the age of 46, long before either he or his tractors had reached their full potential. After his death, development work on the tractors slowed, and, without his energy and imagination, the company lost its momentum. Instead of being a leader in design and marketing, the Ivel company made an unsuccessful attempt to import a Hart-Parr tractor before ceasing to trade in about 1915.

GROWING INFLUENCE IN EUROPE

Although the United States and Britain dominated the early stages of tractor development, there was some activity elsewhere in Europe. Nicholas Cugnot, a French farmer's son, is credited with building the first self-propelled vehicle, using steam power, almost 250 years ago, while Otto built the first successful internal combustion engine in Germany. The French and the Germans used their early leadership, however, to develop motor cars instead of tractors.

A tractor designed and built by a Frenchman called Gougis in about

HORNSBY-AKROYD

Manufacturer: Richard Hornsby & Sons
Model: Patent Safety Oil Traction Engine
Production started: 1896
Power unit: Semi-diesel with blow lamp starting
Power output: 16, 20, 25 and 32HP versions advertised
Transmission: Gear drive with three forward gears and one reverse

1907 included a power take-off shaft to power-trailed machines. Gougis successfully demonstrated the drive shaft with a binder, but it seems that he did not attempt to develop his idea commercially. This was just three years after Professor Scott had shown his p-t-o–driven, front-mounted binder in England, and the idea did not become widely available until the early 1920s.

German interest was concentrated initially on using electricity to power field work such as ploughing, but one exception to this was the ploughing tractor *Pfluglokomotive* designed by Deutz, a company as old as the four-stroke engine. Deutz built two different ploughing tractors in 1907, and one of

Richard Hornsby's Patent Safety Oil Traction Engine was built in small numbers from 1896, powered by a semi-diesel engine. It was designed for stationary and heavy haulage work. A Hornsby became the first tractor to be sold in Britain and another was the first tractor imported to Australia.

these appears to have been the first two-way or bi-directional tractor.

The *Pfluglokomotive* featured an upright steering wheel and control levers at the centre of the tractor, with a seat on each side allowing the driver to face either forwards or to the rear while operating the controls. There were four-furrow ploughs mounted at the front and the rear of the tractor, each with its own cable-operated lift system operated by hand levers. The tractor unit was driven to and fro across the field by using each of the ploughs and also changing seats alternately. Although the tractor was built by Deutz and was powered by a 40HP Deutz engine, the two-way ploughing equipment was called the System Brey after its inventor.

Although the Deutz tractor and ploughing system showed considerable ingenuity, it failed to develop commercially. The Deutz company later became known as Deutz-Fahr, for many years Germany's biggest tractor and machinery manufacturer. It was later taken over by the Italian-based

Same tractor company, now called Same Deutz-Fahr.

A LASTING FUTURE

While the steam engine started the power farming revolution, it was the tractor power pioneered in the United States and Europe that achieved a lasting impact on the structure and economics of food production. The scale of this achievement is indicated by the decline in the number of animals

working on farms. Census figures for the United States show the number of horses and mules on farms peaked at 26 million in 1920, when the number of tractors was almost 250,000. From 1920 onwards, the census figures for working animals were lower each year, reaching fewer than 7.8 million in 1950, while tractor numbers had moved steadily upwards to 3.6 million in the same year.

There were similar trends in Canada. The cultivated area in the prairie provinces had reached 17.4 million hectares (43 million acres) in 1921, the number of working horses had reached 2.24 million – an all-time-high – and there were exactly 38,465 tractors. Thirty years later, the cropped area was more than 27.1 million hectares (67 million acres), the working horse population was 696,000 and farmers used the power provided by a fleet of 237,000 tractors.

Tractor power achieved similar results on British farms, reducing the number

The Ivel established the idea of lightweight, versatile tractors suitable for smaller acreages commercially. The large tank beside the driver's seat holds cooling water for the engine and also provides additional weight over the driving wheels to encourage better traction.

of working horses from their 1,137,000 peak in 1910 to just a few hundred by the late 1990s, when they were mainly used in small numbers for forestry work.

THE FORDSON INFLUENCE

It was Henry Ford's childhood on the family farm in Michigan that sparked his interest in tractor development, and the result was the spectacularly successful Fordson Model F. Other successes from the tractor industry's early years include the lightweight Ivel, the versatile Farmall from International Harvester, crawler tracks from Holt, Best and Caterpillar, and the horizontal two-cylinder engine from the Waterloo Boy and John Deere tractors.

Henry Ford and the Model F had an enormous influence on the tractor industry. The Model F was a basically good design which set many of the standards other tractor manufacturers were forced to follow. Henry Ford's mass-production techniques and his determination to sell the Model F at the lowest possible price made tractor power affordable to many thousands of farmers for the first time and, incidentally, forced many of his rival manufacturers out of business.

Henry Ford was a farmer's son who disliked the slow pace of working the land with horses. He is said to have expressed an interest in building tractors in about 1905, at a time when his car-manufacturing business was on the threshold of becoming one of the biggest success stories in commercial history. The earliest evidence we have of his ideas about tractor design date from 1907, when his first experimental tractor was completed.

When Fordson production was transferred from America to Ireland in 1929, Henry Ford's Model F was updated and the new version was called the Model N. The original sober grey paintwork was eventually replaced with orange and then green.

The first of a long series of experimental tractors designed by Henry Ford and his engineers consists mainly of components from Ford's car production lines, as Ford developed his ideas for cutting production costs. His first tractor had no brakes, but it did feature a leaf-spring front suspension.

The tractor, which has survived and is in the collections of the Henry Ford Museum and Greenfield Village at Dearborn, was built by a team of engineers; it has been suggested that the innovative ideas were theirs rather than Ford's. The flaw in this theory is that it is inconceivable that Ford, well known for his hands-on approach, would not have laid down firm guidelines for his engineers to follow, and he remained closely involved in every stage in the development of his first tractor.

The detail of the design probably came from the engineers, but the innovation is in the overall concept of a small, light tractor designed for mass production and including car components to help reduce the manufacturing cost. This must have come directly from Henry Ford, and it was far removed from conventional thinking in an American tractor industry that was concentrating on building small numbers of large, expensive tractors that only the biggest farms could afford.

FORD'S 'AUTOMOTIVE PLOW'

Henry Ford called the tractor his 'automotive plow', and, when it was complete, it was taken to the Ford family farm for testing. A surviving photograph shows Ford at the controls as the tractor pulls a set of discs in a field where he may once have worked with horses. This was not a photo opportunity suggested by his publicity department – there are patches of soil on Ford's clothing and he looks relaxed.

The tractor was powered by the petrol engine from a Model B car, an up-market model in the Ford range. The engine, which was transversely mounted, had four cylinders, each with its own water jacket for cooling, and the power output was 24HP. The Model B production line also contributed the planetary gearbox, but the tractor's steering gear was from a Model K car, the replacement for the Model B, and the front wheels may have come from the same source. The rear wheels were from an old binder and the traditionally shaped metal seat also appears to have agricultural origins.

A surprising feature of the first Ford tractor is the leaf-spring suspension above the front axle. Front-axle suspension systems were becoming popular by the end of the twentieth century, but it is not clear whether the spring on Ford's first tractors was a far-sighted concession to driver comfort or, more likely, a result of the front axle's motor car ancestry.

More experimental prototypes followed in a variety of shapes, but all were based on the original concept of compact size and sharing components with the car business, and surviving photographs show that many of them

The company was established in 1915, operating from headquarters in Dearborn and owned by Ford and members of his immediate family, and it was called Henry Ford & Son, Inc. This is important, as it explains why the first production tractor was a Fordson, rather than a Ford, product. A few years later, in 1919, Henry Ford was able to buy out the external shareholders in his car company, leaving the Ford family in total control of what at that time was the world's biggest manufacturing business.

THE 'FRAMELESS' TRACTOR

Meanwhile, tractor development was progressing, and a major breakthrough came in 1915 when a new prototype model with 'frameless' construction was completed. The frame was a feature of almost all the early tractor designs, usually made with steel girders to provide a structure to which the engine, transmission and other components could be attached. The frame is easily visible on tractors such as the Waterloo Boy. The Happy Farmer Model G tractor made by the La Crosse Tractor Co. of La Crosse, Wisconsin, also has a clearly visible frame which has the unusual distinction of being made from a single length of steel tube.

included some Model T car parts. The Model T, produced in rapidly growing numbers from 1908, was extremely profitable and provided plenty of financial and other resources for the Ford tractor development programme. More one-off experimental tractors were built and tested – about 50 of them according to one estimate – but Henry Ford was under no financial pressure to move to the production stage, and he preferred to concentrate on making sure that the design was right.

Another factor that may have contributed to the delay in starting the

tractor business was opposition from other shareholders. At that stage, the Ford Motor Company was controlled by a small group of shareholders who had provided the capital Ford needed to start his company in 1903, and they had the power to veto ideas that failed to meet with their approval. The shareholders, who were sharing in the enormous profits from the Model T Ford, were apparently not interested in diversifying into tractor production.

To overcome the problem, Henry Ford decided to set up a separate company to build the new tractor he was planning.

These tractors and many more based on a steel frame structure have plenty of space between the engine and other units, making it easy to see the frame. The spacing also allows easier access for servicing. The disadvantages of an open frame include the problem of protecting sensitive components such as gear

The Wallis Cub was the first tractor to feature the patented U-frame design developed by the Wallis company and introduced in 1913. The curved boiler plate steel frame provided massive structural strength to the tractor, and it also enclosed the underside of the engine and gearbox.

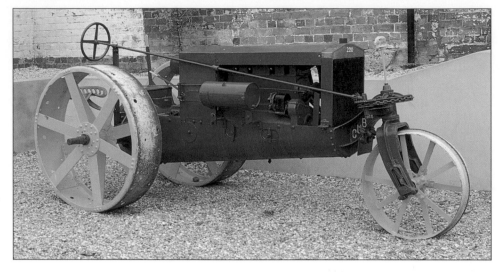

wheels from dust and dirt, and the small amount of flexing that is almost impossible to avoid in this type of frame can put transmission components out of alignment and under stress.

THE WALLIS CUB TRACTOR

The 1915 experimental Ford was not the first tractor to appear with an alternative to the traditional open frame design. Several different ideas had been tried, and the most successful of these was the Wallis Cub tractor with its U-frame made of thick boiler-plate steel. The U-shape provided great strength and rigidity, and it also made it easier to enclose the underside of the transmission, thus protecting it from contamination by dirt and water. The Wallis U-frame made

its first appearance when the Cub tractor was introduced in 1913, at that stage enclosing the underside of the engine and gearbox only. Later versions, however, were extended to include the final drive as well.

Wallis's U-frame, developed by Robert Hendrickson and Clarence Eason in the company's engineering department, was a major step forward in design, and the Cub was a popular small tractor. The

frame remained a feature of the Wallis range and continued to be used after the 1919 merger that saw the Wallis company join forces with the J. I. Case Plow Works.

This merger provided one of the most confusing episodes in tractor history. J. I. Case Plow Works actually had no connection with the J. I. Case Threshing Machine Co., famous for its threshers and steam engines and, later, for

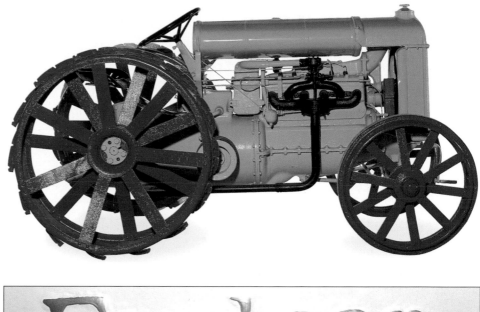

Pressure from the British Government helped to persuade Henry Ford to start building his Model F Fordson in 1917, even though he wanted more time to improve the design. The result was the most successful, and in many ways the most influential, tractor ever built.

tractors, even though both companies were named after Jerome Increase Case and both had their headquarters in Racine. The similar-sounding names continued to cause confusion until Massey-Harris bought the J. I. Case Plow Works in 1929. It paid $2.9 million for the Plow Works and then promptly recovered $700,000 of its investment by selling its rights to the Case trade name to the other Case company.

The distinctive Wallis U-frame was probably one of the factors contributing to Massey-Harris's decision to buy Case, and the company continued to feature it on Massey-Harris tractors until the early 1940s.

Henry Ford's engineering team chose a completely different approach when they developed their frameless design. They joined the engine block, the gearbox and the differential housing together to form one totally rigid, sealed structure that gave a combination of strength and protection. The unit was made of cast iron and was designed to suit the Ford company's expertise in mass-production techniques.

TESTING IN THE FIELD
About 50 pre-production tractors based on the frameless design were built in 1915–16 to take part in a large-scale field test programme. Henry Ford was farming a large area of land he had bought near Dearborn, and this provided plenty of opportunities to test the tractors under commercial conditions. Photographs taken at the time show the test programme tractors bore

A sectioned view of a Massey-Harris 25 tractor equipped with the Wallis U-frame. The 25 arrived in 1931, just three years after Massey-Harris had bought the production rights to the Wallis tractor. The MH 25 was based on the old Wallis 20-30 model, but with changes to the engine.

FORDSON MODEL F

Manufacturer: Henry Ford & Son
Model: Fordson Model F
Production started: 1917
Power unit: Four-cylinder liquid-cooled with 101.6 x 127mm (4 x 5 in) cylinders
Power output: 20HP
Transmission: Fully enclosed gearbox with three forward gears and one reverse

---

some resemblance to the Model F production tractor that was to follow later. The steering wheel was offset to the right on the 1916 tractors, however, and the name on the circular radiator badge on some of them bore the words 'Henry Ford & Son', later shortened to 'Fordson', while others had a blank oval badge with no lettering.

Henry Ford was still not satisfied with the tractors and wanted more design changes, but at that stage the British Government stepped in with a request for an immediate start to production. The war in Europe had reached a critical stage, and large numbers of men and horses had left farms in Britain for the battlefields in Belgium and France. At the same time, German U-boats were destroying so many ships bringing in urgently needed food that there was a serious risk Britain might be starved into submission. Boosting production from Britain's farms had become one of the top priorities, and tractor power was the obvious way to increase output and efficiency. British tractor companies were either working at maximum production levels or had been switched to making military equipment, and some American tractors were already being sold in Britain.

THE MODEL F FORDSON

British officials were aware of Henry Ford's new tractor development project, and this appeared to be exactly what was needed. A compact tractor would be ideal for the small fields, and a lightweight design would offer benefits on the heavy soils of many British farms. The biggest attraction was that Ford planned to mass-produce his new tractor, and this suggested that large numbers could soon be available.

Two of the pre-production tractors were shipped to Britain in April 1917 for testing, and, apart from commenting on a wheel design fault that the Ford engineers had already identified, the

A sectioned view of a 1938 British-built Fordson Model N showing the worm and wormwheel final drive, the air cleaner that replaced the old water-washer cleaner, and the four-cylinder engine descended directly from the original Model F power unit. Apart from a wartime colour change, this version of the Fordson remained unchanged until the E27N arrived in 1945.

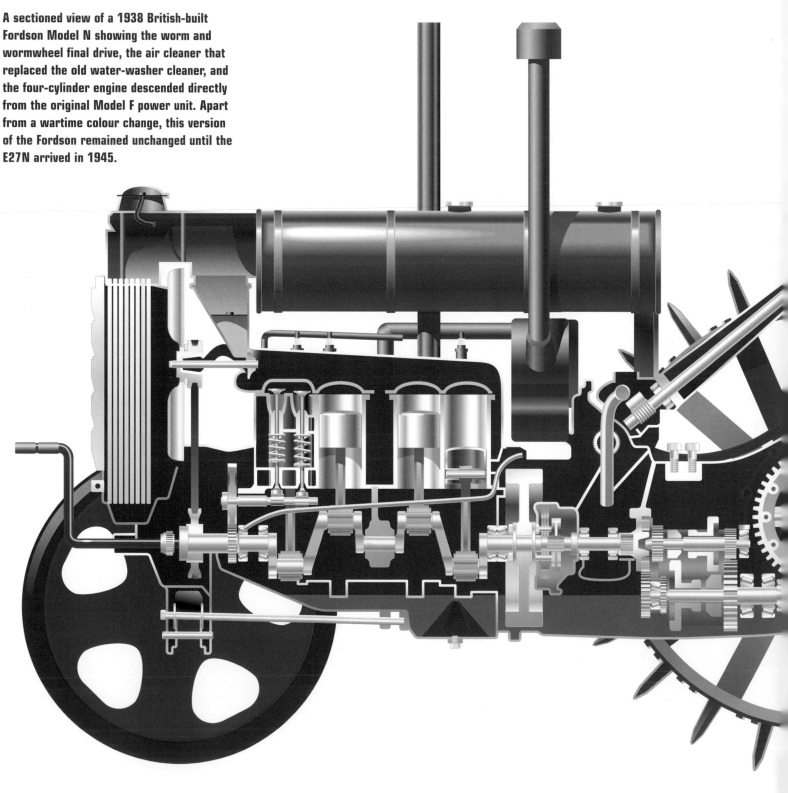

report was highly favourable. A British Government request for tractor production to start as soon as possible soon followed, backed by an order for 6000 of the tractors. Although Ford still wanted more time for development work, he agreed to the British request; on 8 October 1917, the first of the new Model F Fordsons rolled off the production line at the Ford factory in Dearborn.

After a slow start, production increased rapidly, and the British order for 6000 tractors was completed during the spring of 1918. Production peaked at more than 100,000 in both 1923 and 1925, helped by massive orders from the Soviet Government, which depended heavily on American technology for an urgently needed farm mechanization

programme. Soviet imports from the Dearborn factory totalled 26,000 tractors, and thousands of Model F copies were built in the Krasny Putilowitz factory established in Russia with American and British equipment. In 1927, it was estimated that 85 per cent of the tractors and trucks in the Soviet Union were either imported Fords or locally built versions.

Henry Ford was proud of the contribution his tractors were making to increased food production in the Soviet Union, and he also sold large numbers of trucks there to improve transport and distribution. The financial terms he agreed with the Soviets appear to have been generous, and there is some evidence that he lost money on his trade with the Soviet Union. Ford, who was a pacifist as well as being one of the world's wealthiest capitalists, genuinely wanted to provide assistance to the world's first communist state. He felt that, if the ordinary Soviet people enjoyed a good standard of living and were well fed, they would be less willing to go to war.

A GROWING MARKET

The Soviets were not the only customers who were offered cheap tractor power. Ford wanted his tractors to be available to as many farmers as possible, and, instead of trying to maximize his profits, he cut the price of the Model F repeatedly. The list price in America started at $750 in 1918, but it had been reduced to $395 by 1922. The British price was £280 in 1919, but in 1931 the Irish-built Model N version was on the price list at £156.

Although the aggressive price-cutting policy helped to boost Model F sales, it made life difficult for competitors at a time when the market was sliding towards recession. Dozens of tractor companies that had emerged during the

FORDSON MODEL N

Manufacturer: Ford Motor Company
Model: Model N
Production started: 1929
Power unit: Four-cylinder engine. A later version had an oilbath air cleaner
Power output: 27.7HP
Transmission: Three forward gears with an 8.3km/h (5.13mph) top speed

Motor ploughs were fashionable for a few years until they became victims of the Henry Ford's Model F price-cutting policies in the early 1920s. This is the Moline Universal, the most popular of the US-built motor ploughs and the one with the most advanced design features.

boom years of World War I were forced out of business again in the harsher trading conditions of the early 1920s. Fordson sales, meanwhile, soared.

In spite of bargain prices, new design features and its sales success, Henry Ford's Model F tractor was far from perfect, as the Nebraska test figures for 1920 show. In that year, the test centre dealt with more than 60 tractors, including the Fordson, which weighed 1229kg (2710lb), one of the four lightest tractors in the 1920 list. The performance in the belt tests was good, with the Fordson producing more than 19HP in the maximum load test, just ahead of the 18HP rated output, and the fuel efficiency figures from the four-cylinder paraffin burning engine were above average.

The drawbar pull performance was far from good, however, with the Model F recording 23.8 per cent wheelslip, the worst figure in the 1920 test series. As a result of this, the fuel efficiency figure for the drawbar tests was also the worst recorded. High wheelslip figures were also recorded in Britain when two Model Fs were included in a group of tractors taking part in the Society of Motor Manufacturers and Traders (SMMT) trials in 1919. The Fordsons performed reasonably well in good soil conditions where there was good traction, but the wheelslip problem emerged when the tests moved to heavy clay soils, suggesting a lack of weight on the drive wheels or, perhaps, that the lugs or cleats on the wheels were inadequate.

As well as traction tests, the SMMT trials also included a road haulage section, but the two Fordsons were the only tractors excluded from this test. This was because the organizers had decided that the complete absence of brakes made the Model F tractors unsafe for road work.

THE IRISH EXPERIMENT

In 1928, it was decided to end the production of the Model F tractor in the United States and transfer production of an updated version to Ireland. By this time, total production of the Model F was approaching the 750,000 mark, making it the best-selling tractor ever.

Henry Ford's grandfather had emigrated from County Cork in Ireland to find a more prosperous life in the United States, and his grandson had always wanted to set up a factory there to provide jobs in what was then an area of high unemployment. A small number of Model F tractors had been built at a Ford plant in Cork between 1919 and 1922. Closing the US tractor operation in 1928 offered another opportunity to invest in Ireland. Equipment was transferred from the Dearborn factory to the much smaller Ford plant near Cork during the winter of 1928-29. Production started there early in 1929 to meet a backlog of orders that had built up during the production gap, and there was also a big demand for spare parts for existing Fordsons.

The new tractor was known as the Model N and was based on the original Model F design, but there were a number of improvements. Power output from the engine was increased by adding an extra 3.175mm (0.125in) to the bore of the 102 x 127mm (4 x 5in) cylinders, raising the Nebraska power figure from about 19HP to 23.24HP on the belt for the paraffin engine. Heavier, stronger front wheels with five distinctive holes with a rounded triangular pattern were introduced, and the old straight front axle was replaced by a new design with a slight bend in the middle. The Irish-built tractors were also identified with the words 'Irish Free State' or 'Ireland' appearing on the fuel tank end plate.

In commercial terms, the Irish venture was not a success, partly because all the raw materials had to be imported and virtually all the finished tractors were exported; however, the problems were also aggravated by a downturn in the tractor market worldwide. Fordson production was moved once more, this time to the rapidly expanding Ford site at Dagenham, Essex, in England. This was another opportunity to give the ageing design a facelift. Production in England started in 1933 with the old grey paint colour replaced by dark blue, the Fordson name now cast into the radiator side panels, and the radiator top tank carrying a pattern of vertical ribs as well as the Fordson name. This time the fuel tank end plate bore the words 'Ford Motor Company Ltd. England, Made in England'.

More changes came in 1937. A row-crop version called the All-Around was developed belatedly for the American market. Bright orange paint replaced the dark blue, and the old water-washer air cleaner was replaced by an oilbath unit, which was more efficient and also avoided frost damage problems. Another modification was an increase in the compression ratio, designed to increase power output again and also to suit the new higher-octane, paraffin-based fuels that were becoming popular in Britain. The engine changes resulted in several reliability problems; these were rectified by further modifications in 1939.

Another paint colour change – this time to green – came towards the end of 1939 as World War II was looming. Production at the Dagenham plant was increased to provide additional tractors to boost British food production during the war, just as the original Fordson had helped to increase farming productivity at the end of World War I. Despite wartime problems which included shortages of raw materials and bombing raids on the Dagenham factory, the Model N Fordson accounted for more than 90 per cent of Britain's tractor production during World War II. An estimated 80 per cent of the tractors working on British farms at that time were Fordsons.

A NEW FORDSON

The last chapter in the development of the Model F and its descendants came in 1945, with the launch of the new Fordson E27N. There was the inevitable colour change, this time to dark blue, but the other improvements were more comprehensive. The new model was bigger and heavier than its predecessor, and the expensive, power-consuming worm and worm wheel final drive of the Models F and N were replaced.

Victims of the price-cutting campaign that contributed to the Fordson sales success in the early 1920s included the motor plough, offered as a low-cost alternative to the conventional tractor. Motor ploughs were popular in the United States – and to a lesser extent in Europe – from about 1915 until the Fordson ended their brief success story

The Moline Plow Co. had been building ploughs and other implements since the 1860s; in 1915, they took over Universal Tractor Co. and began making the Universal motor plough. Following a sales slump, in 1929, Moline Plow merged with two other companies to form Minneapolis-Moline Co.

in the early 1920s. The design included what was usually a small engine driving a pair of wheels near the front, with a frame extending to the rear to allow an implement such as a plough or interrow hoe to be attached under the frame. The rear end of the frame was commonly supported by a small pair of wheels, with the driving seat attached above the frame; the most basic models were pedestrian controlled and had no driver's seat.

Motor ploughs were often underpowered and difficult to steer accurately, but the idea had a few definite advantages. Placing the driver right at the rear ensured an excellent view of interrow equipment such as hoes and weeders, and this made motor ploughs popular in America for rowcrop work. The biggest attraction, however, was relatively low prices for the tractor unit and the small implements they were designed to operate.

THE MOTOR PLOUGH

America's leading motor plough was the Universal made by the Moline Plow Co. of Moline, Illinois. It was introduced originally in 1914 with a 12HP petrol engine with two horizontally opposed cylinders. This model was updated in about 1918 by the Model D version, which had a four-cylinder vertical engine replacing the original power unit, boosting the output to 18HP. An unusual feature was the electrical system based on a battery and including lights and a starter motor, making the Universal Model D the first tractor available with electric starting. In 1929, the Moline Plow Co. merged with two other companies to form the Minneapolis-Moline Co.

Allis-Chalmers moved into the motor plough business in 1919 with its 6-12 model, powered by a LeRoi 2C engine rated at 15.6HP. One indication of the sales slump that hit the motor plough market is the change in the list price of the 6-12. The price in 1919 when the tractor was launched was $850, but a series of reductions brought this figure down to only $295 in 1923, when Allis-Chalmers was trying to find customers for the remaining unsold stock of 6-12s.

The motor plough fashion was already past its peak by 1920, when Chandler & Taylor took out the patents for their Adaptable motor plough designed by Mr Chandler. The company was based in Indianapolis, Indiana, and its motor plough was given a 10HP power rating on the drawbar and 20HP on the belt. In spite of the versatility claimed for the design, very few of the tractors were sold.

If there were a prize for the most unconventional motor plough design, the Bates Steel Mule would certainly be on the short list. The Bates Machine and Tractor Co. of Joliet, Illinois, announced the 13-30 model in about 1915. It used a single crawler track at the rear of the main tractor unit to provide the traction and two wheels at the front for steering. Implements were hitched to the rear, and the operator sat at the rear on a seat attached to the implement. This arrangement put the driver a long way behind the tractor unit, although the exact distance varied with the size of the implement. To cope with a range of distances, the steering column and other important controls were all extendable. Although the 13-30 Steel Mule looked quite strange, it was surprisingly popular, but the company later switched to more conventional tractor designs and continued building half-track, full track and wheeled models until 1937.

British interest in motor ploughs failed to reach American levels of enthusiasm, but one model which did achieve limited success during World War I was the pedestrian-controlled Crawley Agrimotor. Designed by a farming family in Essex, England, the prototype version was built for the Crawley farm in 1908; it attracted sufficient interest for the Garret company of Leiston, Suffolk, to agree to build a small batch of the tractors. Production was later taken over by the Crawley family, using a Buda engine with a 30HP output.

Fowler, the Leeds-based agricultural steam engine manufacturer, decided to diversify into the motor plough market in 1923, when interest in the idea was already well past its peak. It chose an Australian design, which it built under licence and called the 'Rein Control' motor plough. It included a set of reins to operate all the controls, an arrangement that presumably would have

Massive chain and sprocket drive systems were the usual method of transmitting power from the engine to the rear wheels until gear drives took over. It was reasonably reliable, but as this International Harvester Mogul 8-16 tractor shows, the system was completely exposed to dirt and dust.

International Harvester scored a big success when it announced the Titan 10-20 in 1916. The design was traditional, with the engine and other components mounted individually on the steel girder frame, but the Titan was mechanically simple and easy to service, and it earned a reputation for long-term reliability. Production ended in 1922 after more than 78,000 of the 10-20 tractors had been sold.

International Harvester's Farmall tractor introduced important new design features when production started in 1924. It offered significant advantages for rowcrop work, and it was so successful that most of the major American tractor companies introduced their own versions.

seemed reassuringly familiar to anyone accustomed to working with horses. A sharp tug on the left or right rein operated the steering, a firm, steady pull on both reins brought the tractor to a stop – just like driving a horse – and a second firm pull engaged the reverse gear. Several tractor and motor plough manufacturers in the United States had used a similar system, but there, as in Britain, demand was small.

While most American and British manufacturers treated the motor plough as an economy machine, some of the motor plough developments in other European countries were much more ambitious. The Praga company in what is now the Czech Republic was building motor ploughs from about 1910, with the production total reaching more than 1500 by the mid-1920s. The power unit was a four-cylinder petrol engine producing 45HP.

Another Czech-based company, Laurin and Klement, was also concentrating on motor plough production. It was the forerunner of the Skoda car company, and its motor ploughs were sold under the Excelsior brand name. Two models were available, with

engines producing 40HP and 50HP, respectively, and they were popular for about 15 years until production finished in about 1925. High horsepower motor ploughs were also built in Germany. The Austrian company Puch – which later became part of the Steyr Daimler Puch company building Steyr tractors – developed a motor plough powered by a 7.0-litre (427-cu. in) petrol engine. The engine produced 40HP, and the plough's driving wheels were more than 1.83m (6ft) in diameter.

The most important factor in the ending of the fashion for the motor plough was the success of the Fordson Model F and Henry Ford's price-cutting policy. Motor ploughs relied on low prices to attract customers, but the

FARMALL

Manufacturer: International Harvester
Model: Farmall
Production started: 1924
Power unit: Four-cylinder valve in head engine burning paraffin
Power output: 18HP
Transmission: Enclosed gearbox with three forward speeds

In 1906, tractors were added to the highly successful range of general farm machinery and stationary engines; within five years, International Harvester had taken the lead as the largest farm equipment company in the world.

World War II provided further opportunities for expansion, and the International Harvester success story continued with the highly successful Mogul tractor range sold through McCormick dealers and the Titan tractors available from Deering dealers. The Moguls and Titans were conventional in design and layout, but they were ruggedly built. The engines that provided the power for both ranges had established a reputation for reliability, and International Harvester continued to lead the industry. The list of models also included the Junior 8-16, a lightweight tractor with an unusual layout. The vertical radiator was mounted just behind the engine and in front of the steering column and cylindrical fuel tank, which allowed a distinctive bonnet line sloping sharply down at the front to give excellent forward visibility from the driver's seat.

International Harvester, the established market leader offering a range of traditional and somewhat dated Titans and Moguls, was especially vulnerable to the success of the cheap and cheerful Fordson. The management realized that the company's survival as a major force in the tractor market depended on bringing in some successful new models. One of the ideas apparently favoured by the engineers was developing a steam tractor with up-to-date technology. This venture, described in 'Power Alternatives, was almost certainly a dead end in commercial terms, and it was abandoned.

The tractor that played a crucial role in International's success during the 1920s and 1930s was the Farmall, first available in 1924, with frameless construction replacing the traditional layout of the Moguls and Titans. The feature which brought well-deserved success to the Farmall, however, was the fact that it was designed to meet the needs of rowcrop farmers. The new tractor was, of course, capable of doing other types of farm work as well – which is why the Farmall name was appropriate – but it offered special advantages on farms where interrow jobs such as hoeing

Fordson was just as cheap to buy, and it did almost every job more effectively than a motor plough. The exception was rowcrop work, where the driving position and the high clearance available from most of the American-designed motor ploughs gave them a continuing advantage over ordinary tractors.

INTERNATIONAL HARVESTER

Even the rowcrop advantage was soon to disappear. While many tractor companies were forced out of business or disappeared in mergers or as the victims of takeovers, some were able to fight back successfully, and these included International Harvester. International Harvester was formed in 1902, when Deering and McCormick joined forces.

FRAME OR FRAMELESS

The frameless design of the Model F Fordson was a major step forward in tractor development, and other manufacturers were quick to introduce new models to replace their old steel girder frame designs.

But some manufacturers were reluctant to abandon the framed design, and the list includes most of the leading crawler tractor manufacturers. They adopted the Fordson idea of linking the engine and transmission to provide a rigid, dirt proof structure, but they also retained the frame to give extra structural strength in tractors used mainly for heavy draft work.

The number of manufacturers using a frame plus an engine/transmission structure has recently increased. This follows the big increase in engine power since the early 1990s, and for tractors in the 200 to 300HP sector manufacturers such as Case IH and Deere say adding a frame helps to provide the strength and rigidity needed to cope with the stresses imposed when a big tractor is working.

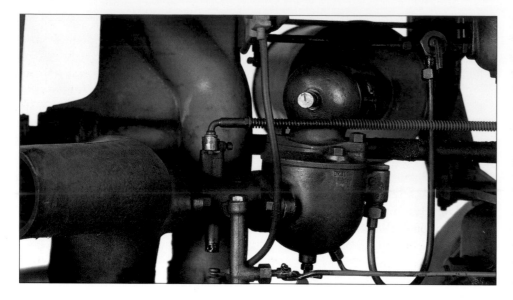

This 1926 John Deere Model D tractor features the original Schebler fuel primer and carburettor. The Model D was the first tractor designed by John Deere after the Waterloo Boy takeover. There is a strong demand for the earlier versions among the many thousands of John Deere enthusiasts.

models from other manufacturers, but it was International Harvester that brought it all together in the Farmall, and they reaped the rewards.

A pre-production batch of Farmalls was completed in 1922, and most of these were sent to farms in Texas for evaluation. The reports were enthusiastic,

were important. It was this that put the final nail in the motor plough's coffin.

Crops grown in rows, such as cotton and maize or corn, occupy vast areas of farming land in the United States, and a tractor that made a genuine contribution to efficiency for this type of farming was virtually assured of success. Development work on the rowcrop tractor project began in 1920, and Bert R. Benjamin of the International Harvester engineering department in Chicago has been credited with the Farmall design. One important feature was the fact that there was plenty of space under the engine and transmission to allow the tractor to straddle the rows without damaging the crop plants. The generous clearance also allowed space for mounting implements such as hoes and ridgers under the middle of the tractor between the front and rear wheels. Rowcrop work demands accurate steering to destroy weeds effectively, while avoiding damage to crop plants; mid-mounting allows a good view of the implement from the driver's seat.

Other design features to attract rowcrop farmers included relatively light weight to minimize soil damage, plus a brake control for the steering which allowed the tractor to pivot on either of the rear wheels to make headland turns sharper. Later versions of the Farmall also provided a generous range of adjustment for the wheel spacing to suit different row widths. Some of these features had appeared previously on

A cutaway of a Model D clearly shows the mechanical simplicity that helped to make this and later two-cylinder John Deere tractors such a success. The heavy-duty transmission with only two forward speeds and a reverse included a rugged chain and sprocket drive to the rear wheels.

JOHN DEERE D

Manufacturer: Deere & Co.
Model: Model D
Production started: 1923
Power unit: Two-cylinder horizontal engine with 165mm (6.5in) bore
Power output: 30HP maximum
Transmission: Two forward speeds

and production started in 1924, using a four-cylinder IH engine delivering 18HP at the belt pulley in its Nebraska test. This was about the same as the Fordson, but the advantages of the special design features of the Farmall allowed International Harvester to sell it successfully with a list price of $950 for the tractor plus an extra $88.50 for the mid-mounted cultivator, more than double the Fordson price.

More Farmall models followed. The F20 version, based on the previous model but with a 10 per cent increase in engine output to 20HP, arrived in 1931, together with the 32HP F30 Farmall.

Another new model arrived in 1932, when the F12 was announced with a 16HP engine, to be followed by the two-plough F14 Farmall in 1938. The success of the Farmall encouraged most of the leading US tractor makers into the row-crop market. Deere's contribution was the GP or General Purpose tractor, Case offered a special rowcrop version of its Model C, Huber introduced the LC, and Minneapolis-Moline designed its Twin City Universal model for rowcrops.

International also continued to make more conventional tractors in large numbers during the 1920s, replacing the old Titans and Moguls with more up-

to-date designs based on a one-piece front-to-rear frame. The first of the new generation tractors was the 15-30 Gear Drive model, available from 1921 and joined in 1923 by the smaller 10-20 Gear Drive model. Both tractors inherited the sturdy design and construction of the previous models, and both were available with some design improvements into the early 1930s.

JOHN DEERE

Another US company that survived the 1920s Fordson competition without, it seems, too many problems was Deere & Co. It continued to sell the Waterloo

Henry Ford insisted on building just one version of the Model F to simplify production and keep the selling price low, but other companies recognized the opportunity to produce conversion kits. One was the Full-Crawler Co., which produced a tracklaying version.

of the General Purpose rowcrop tractor, arrived in 1927, and production of the Model A started in 1934, followed by the Model B in 1935 – all with two-cylinder engines, of course.

THE CRAWLER TRACTOR

Crawler tractors were another sector of the market suffering relatively little damage from the cut-price Fordson. Henry Ford's production economics depended on a high degree of standardization, and he never produced a rowcrop or a crawler version of his Model F tractor – although some tracklayer conversions were available from other companies. These included the Trackson conversion available from the Full-Crawler Co. of Milwaukee, Wisconsin, during the mid-1920s. The Trackson replaced both the front and rear wheels to convert the Model F into a full-track crawler tractor.

Four-wheel drive was not widely available in the 1920s and 1930s, and, for those who wanted extra pulling efficiency, a crawler tractor was the obvious option. Crawler tracks have always been the most effective way to convert engine power into drawbar pull, particularly in wet soil conditions when steel cleats or the lugs on a modern rubber tyre do not grip efficiently. But tracks also offer other benefits, including a bigger contact area on the ground to spread the weight of the tractor and reduce the risk of surface ruts and compacted soil. The extra grip of the tracks also improves stability and safety when working on steep land.

Inventors were working on the idea of using tracks instead of wheels long before the first tractor was built. The first crawler-track patents in Britain can be traced back to the 1770s, and several unsuccessful attempts were made to produce a steam-powered tracklaying vehicle in the early nineteenth century.

Boy Model N during the early 1920s, while it worked on a replacement tractor to be sold under the John Deere brand name. During this time, the Waterloo Boy, increasingly outdated with its steel frame design and exposed drive gears, was attracting fewer than 1000 customers per year. Its replacement, the Model D, was announced in 1923, but Waterloo Boy production continued well into the following year.

The Model D was lighter, more compact and more manoeuvrable than its predecessor. It provided more power, and it was based on a more up-to-date frameless design with enclosed gear drives, but one important design feature inherited from the Waterloo Boy tractors was the twin-cylinder horizontal engine. In fact, the engine on the Model D, and on virtually all other John Deere tractors produced until 1961, was the opposite way round. The cylinder heads were at the front instead of at the rear, as they had been on the Waterloo Boy tractors - but it was still a big, slow-revving two-cylinder engine, one that continued to be a feature of John Deere tractors for almost 40 years.

The fact that the John Deere design team decided to retain the two-cylinder engine is surprising. Most of the other tractor manufacturers in the United States and Europe were switching to

four cylinders – in fact, most of them had already done so – and the two-cylinder engine in the new Model D may have seemed old-fashioned in what was otherwise an up-to-date design.

History has shown that the choice of engine layouts for the Model D was correct and the twin cylinder design was probably an important factor in establishing Deere as one of the leading US tractor manufacturers, particularly during the 1920s and 1930s. The twin-cylinder engine was slower running and not as smooth as its rivals with four cylinders, but smoothness was not a big sales feature in the 1920s, and the two-cylinder engine offered other benefits that were more important.

It was a much more simple design with a reputation for long-term durability, and it was also easier to service and repair, an important advantage at a time when skilled mechanics were few and far between. Two cylinders instead of four meant fewer working parts and the prospect of improved reliability, and this was also a major attraction at a time when many customers were still first-time tractor buyers with little or no previous experience of tractor operation and maintenance. Sales of the Model D had reached the 10,000-per-year level by 1928, and more models were added to the range. The Model C, forerunner

More experiments with tracks on steam vehicles followed in the United States and in Britain, and small numbers had been built and demonstrated during the second half of the century.

One of the first steam-powered tracklayers to achieve a small measure of commercial success was designed by Alvin Lombard of Waterville, Maine. In the 1890s, he was experimenting with tracked vehicles for moving newly felled timber, and he sold a small number of these in the early 1900s. They were based on a half-track design, with front wheels for steering, but he also built a 100HP version equipped with skis instead of wheels at the front for travelling over snow.

The real pioneers of the tracklaying tractor were the Holt and Best companies in the United States and Richard Hornsby in England. David Roberts, chief engineer with the Hornsby company, developed a track system which worked reasonably well, and one of the Hornsby-Akroyd

tractors was modified to run on the tracks in about 1904. It is not clear whether the company was planning to use the new tracklaying tractor for farm work, but we do know that it was aware of the potential benefits of using tracks for a military vehicle. The Hornsby tractor on its Roberts tracks was demonstrated to the British War Office to show how effectively it could deal with difficult ground conditions. The military chiefs remained loyal to their cavalry horses, however, and the Hornsby forerunner of the army tank was given the same thumbs down that Dan Albone had received for his prototype military ambulance based on one of his Ivel tractors (see 'Start of the Power Farming Revolution').

BENJAMIN HOLT AND DANIEL BEST

Not unexpectedly, Richard Hornsby lost interest in crawler tractor development, but the Best and Holt companies in California both concentrated on developing agricultural tracklayers, and

they were much more successful. Benjamin Holt, who had moved from New England to Stockton, California, in 1883, started a wheel-making business. A few years later, he branched out, rather surprisingly, into building some of the first combine harvesters. The Holt company's biggest rival in the combine market was Daniel Best, a farm boy from Iowa who had first settled in Oakland, California, and later moved to San Leandro.

The original combines, which were big and heavy, were pulled by teams of horses or mules with as many as 40 animals in some of the teams. Both Best and Holt built steam traction engines to replace the horses, and both companies

Richard Hornsby spent a great deal of time and money developing a tracked version of his Hornsby-Akroyd tractor. This photograph was taken during official trials, when the British War Office's military chiefs failed to appreciate the potential benefits of this predecessor of the army tank.

Daniel Best, a farm boy from Iowa, started the company that carried his name and made a major contribution to crawler tractor development. The photograph shows a Best 60 tractor, powered by a four-cylinder engine developing 60HP, production of which started in 1919.

the finance acquired from the sale of the old business to start the new one.

The C. L. Best company began building wheeled tractors at first, but crawler tractors were added to the range in 1913, starting with the Model 75 powered by a Best 197 x 229mm (7.75 x 9in) four-cylinder engine, and the Best and Holt businesses were rivals once more. The renewed competition ended in 1925, when the two companies agreed to merge under the Caterpillar brand name and became based at the Caterpillar headquarters at Peoria, From that time onwards, Caterpillar has been the leading crawler tractor manufacturer worldwide.

Following the merger, the new Caterpillar tractor range was established by selecting models from the old Best and Holt ranges. The chosen few included the big 30 and 60 models from the Best stable at the top of the range, plus new 2 Ton, 5 Ton and 10 Ton models from Caterpillar. In 1931, Caterpillar became the first tractor company in the United States to offer a production tractor with a diesel engine (see 'Designed for Performance').

CRAWLER TRACTOR SPECIALISTS

The United States was easily the world's biggest market for crawler tractors in the 1920s and 1930s, and Caterpillar faced plenty of competition, both from tracklayer specialists and from tracked models introduced by some of the big-wheeled tractor companies. The specialists included the Cletrac range from a company established by Rollin H. White, a member of the family that had produced the White steam car.

Rollin White was interested in the tractor industry and in crawler tractors in particular, and he was a member of the group of businessmen who formed the Cleveland Motor Plow Co. based at Cleveland, Ohio, in 1916. This followed

were faced with the problem of using heavy equipment on the soft peat soils in some of the principle grain-growing areas. The solution they both chose was to design extra-wide wheels for the steam engines to spread the load over a bigger surface area, and both companies were apparently in competition to build the biggest and most spectacular sets of wheels. When Best built a traction engine in 1900 with wood-covered driving wheels 2.74m (9ft) high and 4.6m (15ft) wide, the Holt company went one better. Holt's engine had sets of three driving wheels on each side, with a 1.83m (6ft) width for each wheel and an overall width of 5.49m (18ft) on each side. Fortunately, there was little traffic on California roads in the early 1900s, as a slow-moving steam engine with 4.6m (15ft) or 5.49m (18ft) wide wheels on each side could lead to problems on a busy highway.

Ridiculously wide wheels were obviously not a long-term answer to the problem of soft soils, and Benjamin Holt's company built a steam traction engine on tracks in 1904. Holt built a small number of steam-powered track-layers before switching in about 1907 to a crawler tractor powered by a petrol engine of its own design. The tractor was one of the first Holt products to bear the Caterpillar name, which was formally registered in 1910, the year after the company had expanded into a new factory at Peoria, Illinois.

In 1908, the intense rivalry between the two California-based combine and steam engine manufacturers ended temporarily when the Best company was taken over by Holt. The peace proved to be short-lived when C. L. Best, the son of Daniel, started a new company in 1910 called the C. L. Best Gas Traction Co., presumably using at least some of the development of a new controlled differential steering system for crawler tractors. The new system used gears and a steering wheel instead of the usual steering clutches and levers, and it offered a number of advantages for the operator. The steering was controlled by a wheel, just like a conventional tractor, and turning the wheel applied a braking force to the gear drive to the track on one side, while the other track continued to operate under power. This steered the tractor left or right, and was said to give improved accuracy when manoeuvring and positioning.

The first of the Cleveland tractors was the Model R, available from 1916 and powered by an 18HP Weidley engine with four cylinders. The engine, which developed its rated power at 1200 rpm, was mounted transversely, with the belt pulley operating at right angles to the tractor. This arrangement's advantages are not obscure, but the big disadvantage must have been the problem of positioning the tractor correctly to give the right amount of belt tension.

Presumably the Model R tractor needed further refinement, as it was replaced by the improved Model H in 1917, perhaps following the founder's initials. The design changes included engine and track-wheel modifications, but the sideways-facing belt pulley was retained. The company name was changed to the Cleveland Tractor Co. in 1917, and the Cletrac brand name was adopted in 1918. The company continued to be prominent in the tracklayer market until 1944, when it became part of the Oliver group.

Another of the tracklayer specialists was the Monarch Tractor Co. This company was established at Watertown, Wisconsin, just before the start of World War I, and it started with a range of small, lightweight crawler tractors. The early Monarch models included the Lightfoot with a 6-12HP power rating from a small four-cylinder engine, and the Neverslip was rated at 20HP on the belt pulley and 12HP at the drawbar using an Erd four-cylinder engine.

The company appears to have encountered financial problems at various times, as there were several reorganizations, but, in the mid-1920s, it moved sharply up-market with the Model D 6-60. The model number indicated 60HP output from the rare luxury of a six-cylinder Beaver engine. The engine was a valve-in-head design with a 1200 rpm speed rating, and it was linked to a three-speed gearbox. The 6-Ton model followed in 1926, recording 50.55HP at the drawbar when it was tested in 1927 at Nebraska. Monarch's biggest tractor, the 11.7-tonne (11.5-ton) 75 model, was added to the range early in 1928, just before the Monarch company was bought by Allis-Chalmers to add tracklayers to the company's range of wheeled tractors.

OTHER TRACKLAYER MODELS

International Harvester made its first move into the tracklayer market in 1928 with crawler versions of its popular 10-20 and 15-30 models. They were called the International Harvester TracTracTors, and the person responsible for suggesting such a descriptive and easily remembered name deserved a generous pay rise. A new TracTracTor available from 1932 in the McCormick-Deering line was the first International tractor offered with petrol and diesel power options. The petrol-engined TA-40 was powered by a six-cylinder IH engine powering a five-speed gearbox and the TD-40 diesel engine was a four-cylinder design.

John Deere tracklayers were available from 1935, but, as described in 'Postwar Expansion', they started out initially as orchard tractors which were equipped with tracks in a special conversion carried out in Yakima, Washington, by the Lindeman brothers.

US-built crawler tractors dominated the tracklayer market in Britain throughout the 1930s, but there were home-produced models near the top and the bottom of the power range. The small crawler tractor was the Ransomes MG series, with the letters MG standing

One of the special features of the Cletrac range of tracklayers was the steering system, which used a wheel instead of the twin levers favoured by most of the other crawler tractor companies. This was the system used on the Cletrac 20K crawler, a 20HP four-cylinder machine.

for 'market garden'. The first production version of the mini-sized tracklayer arrived in 1936. It was called the MG2 and it was powered by a single-cylinder Sturmey-Archer engine which was air-cooled and developed 6HP. When the replacement model was announced in 1949, it was called the MG5 – no explanation was offered for the absence of the MG3 and 4 models – and the final version was the diesel-powered MG40. The MG drive system included a centrifugal clutch set to disengage at a 500rpm engine speed, and power was delivered to the tracks through a set of reduction gears and two crown wheels, one for forward travel and the other for reverse, and there was also a differential. Each of the steering levers applied a brake on one of the tracks, forcing the differential to speed up the drive to the unbraked track.

MG production was spread over a 30-year period; during that time, the sales total reached approximately 15,000. Most of the tractors were sold to market gardens and smallholdings, but there were also some unexpected markets. They were exported to Tanzania to scrape salt from the surface of inland saltpans, and some were sold to Dutch farmers who found them light enough to be ferried in a small boat across the drainage dykes separating some of their fields.

Easily the biggest of the British-built tracklayers in the 1920s and 1930s was the Gyrotiller, made by the Fowler company in Leeds, Yorkshire. Its success encouraged Fowler to build a range of more conventional tracklaying tractors during the 1930s to compete against US imports. Gyrotiller production ended in 1937, leaving the company once more

short of products and facing serious financial problems. It was the rapidly increasing prospect of war that brought a short-term solution to its problems, as the British Government looked for companies to build equipment for the armed services. Fowler, with its vast factory space, its long and generally successful history in heavy engineering and its recent experience of building tracklaying vehicles, was an obvious candidate to build some of the tanks needed by the British Army, and tank

production filled the Leeds factory throughout most of the war.

THE MILITARY CONNECTION

The possible connection between the Gyrotiller and Fowler's tank production contract was one of many links between crawler tractors and some of the specialist vehicles used in warfare. In spite of their decision to reject the tracked version of the Hornsby- Akroyd tractor for military use in about 1905, the British Army was the first to use

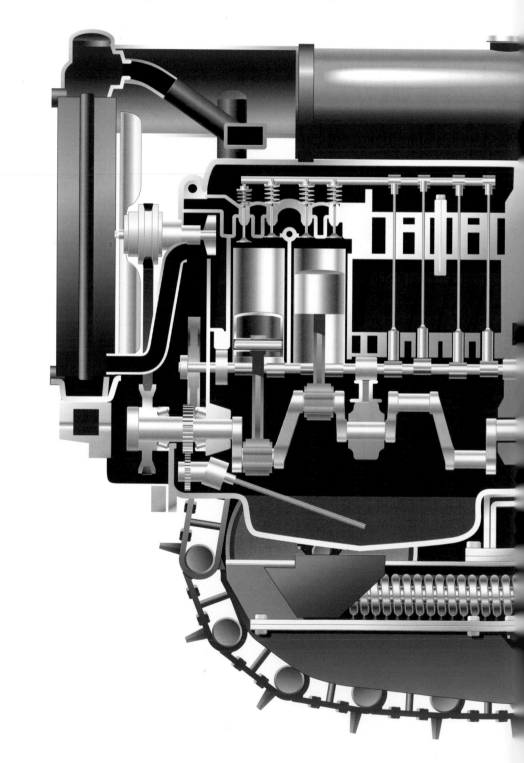

MONARCH M

Manufacturer: Allis-Chalmers
Model: Model M tracklayer
Production started: 1934
Power unit: Four-cylinder Allis-Chalmers Model U engine
Power output: 34HP
Transmission: Gear drive with three forward speeds

tanks for warfare. The first tanks were the result of what the army code-named the Landship Project, a high-priority programme aimed at developing a tracked armoured vehicle that could travel across the mud, shell craters and trench systems of a World War I battlefield.

At an early stage in the war, the British Army had ordered a number of crawler tractors from the Holt company in California, and these were performing well as the towing vehicles

for heavy guns and other military equipment at the battle front. It is not clear whether the Landship Project was prompted by the success of the Holt Caterpillar tractors or by memories of the Hornsby-Akroyd demonstration a few years earlier, but the army engineers looked to the United States for their crawler-track technology.

Several US-built crawler tractors were imported for evaluation, and the list included a Bullock Creeping Grip from Chicago and a Strait's Tractor designed

by William Strait and built by the Killen-Walsh Co. of Appleton, Wisconsin. The Strait's Tractor was an unconventional design built on a tricycle layout of three small tracks, but

Allis-Chalmers moved into the crawler tractor market when it bought the Monarch company, a tracklayer specialist, in 1928. This cutaway picture shows the four-cylinder engine and the transmission to the rear pair of track sprockets. It also shows the cushioned seat provided for the driver.

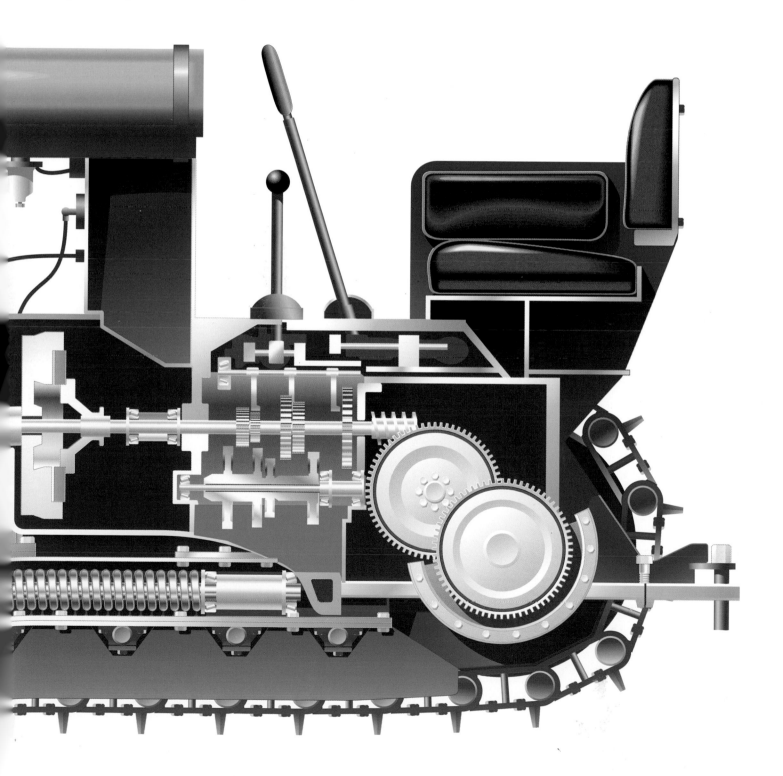

The tiny MG crawler tractor series from Ransomes provided the benefits of tracks for smallholdings and small-scale commercial growers, and it was a success. Production started in 1936 using a 6HP air-cooled engine and rubber-jointed tracks developed by Roadless Traction.

this was the tractor that may have had the most influence on the Landship Project. It was provided with an armoured body shell for protection and was demonstrated in front of senior military and political figures, including the young Winston Churchill. One feature which apparently impressed the Landship committee was the way in which the tractor's main driving tracks tilted upwards at the rear, allowing the tractor to be reversed over obstructions. The tracks on the first British tank and on many of the later designs have a similar tilt-up design feature to give increased mobility.

While tractors contributed to the design of the first tanks, the reverse process also occurred. During World War I, the Renault car company developed a light assault tank for the French Army, and this provided the basis for the GP tracklayer announced in 1919, Renault's first tractor. The GP used the crawler-track technology Renault had built up during the development and production of its tank, and the power unit was a 30HP four-cylinder engine, with the power delivered through a cone clutch and a three-speed transmission.

An unusual feature of the GP tractor was the tiller steering with its two handles arranged like the handlebars of a bicycle. Presumably GP owners liked this arrangement, as it was retained when the GP tractor was updated after

about two years to produce the H1 tracklayer, plus a wheeled version called the H0. Another design feature of the first Renault tractors was the position of the radiator, which was behind, rather than in front of, the engine in order to allow a downward-sloping bonnet line to improve the forward view from the driving seat. A similar arrangement was also used by International Harvester on its 8-16 Junior tractor, and the same radiator position was retained when the H1 and H0 tractors were replaced in 1927 by the new PE model.

Renault's postwar arrival in the crawler tractor market coincided with the launch of the first tractors from its car industry rivals, Citroën and Peugeot. Both companies built tracklayers, but it was Renault that emerged as the biggest and most successful tractor manufacturer in France, a position it still holds. There was also a strong British influence in the early years of the tractor industry in France. The Saunderson Universal, the best-selling British-built tractor in the 10 years or so before 1920, was assembled in France under the Scemia

A small tank developed for the French Army during World War I provided the base for the first Renault tractor, the GP tracklayer. An unusual design feature is the location of the radiator, positioned behind the engine and below the fuel tank to allow a sloping bonnet line for improved forward visibility.

brand name. The Austin tractor, the top-selling British built-tractor during the early 1920s, was later assembled at a factory near Paris, with most of the components imported from the Austin plant in England. When Austin decided to stop making the tractor in Britain to free up additional production space for its cars, British customers were supplied with French-built Austin tractors.

Another example of the swords-to-ploughshares philosophy of converting tracklaying military vehicles for farm

use followed the end of World War II in 1945. Between 1941 and 1944, the NSU company in Germany had built the Ketten-Krad, a powerful motorcycle built on a pair of tracks, with one motorcycle-type wheel at the front for steering. The tracks, one on each side, gave the Ketten-Krad an almost unstoppable crosscountry ability, taking the driver and up to two passengers over mud and snow, while its small size allowed it to use narrow tracks and drive through woodland. It was also light enough to be dropped by parachute from a small aircraft.

The Ketten-Krad's lively performance was provided by a four-cylinder Opel car engine, producing 36HP from 1.5 litres (91.5 cu. in) capacity. The engine was used before and after the war in the Opel Olympia car, and there was a

three-speed gearbox plus a second low-ratio box for off-road driving. Ketten-Krad production totalled 8345, and most of those that survived the war were bought by farmers and foresters for use as personal transport and for carrying light equipment. As a farm vehicle, the NSU Ketten-Krad was an early ancestor of the lightweight, go-anywhere ATV, or all-terrain vehicle, developed during the 1980s by Japan's motorcycle industry and now widely used as a low-cost runabout vehicle on thousands of farms.

Italy's first move into full-scale commercial tractor production followed the end of World War I, when the Fiat company produced the 702 model. Fiat was the principal manufacturer of trucks for the Italian Army during the war; the slogan used to promote the new tractor was 'From machines of war to machines

for the earth'. This was appropriate because the power unit used for the new tractor was inherited from the 18BL truck built by Fiat in large numbers as the army's standard transport vehicle.

A prototype version of the 702 was completed before the end of 1918. It featured a frameless design with the engine and gearbox joined together as a sealed unit, suggesting that Fiat followed very closely behind Ford in developing this feature. The truck engine used in the 702 tractor was a four-cylinder petrol engine producing 30HP, and it was linked to a three-speed gearbox which also provided three speeds on the belt pulley. The 702 was followed by improved versions called the 702B and 703, and the 700A tractor was announced in 1927 and remained in production in various versions until 1949.

DESIGNED FOR PERFORMANCE

Harry Ferguson's three-point linkage with hydraulic control was one of the important technical breakthroughs of the 1920s and 1930s. Other power farming developments from this period included the first inflatable tractor tyres advertised by Model U racing tractors from Allis-Chalmers, and the arrival of diesel power to boost pulling power and reduce fuel costs. Manufacturers also discovered that stylish lines and eye-catching colours help to sell more tractors.

The launch of Henry Ford's Model F Fordson in 1917 was one of the most important events in the history of the tractor – many would say the most important. It was the Model F that helped to establish the layout and structure of the modern tractor, and the Model F's bargain price brought affordable tractor power to many farms for the first time.

Although the Model F was a major milestone in design, there was still plenty of room for further technical improvements to make tractors more productive and efficient, and some of the most important developments were introduced during the 30-year period from 1920. One of the most significant improvements was Harry Ferguson's hydraulically operated three-point linkage with draught control, which sensed changes in the resistance of the soil against the implement to adjust the working depth and brought the tractor and implement together to operate as

This superbly restored John Deere is a Model R, the company's first production tractor to be powered by a diesel engine. It was also the most powerful tractor John Deere had built, delivering up to 51HP, and also set a new record for fuel efficiency.

a single unit. This was another major development in which Henry Ford played a leading role.

The first attempts to pull ploughs and other implements had been based on a hitch point at the rear of the tractor and a drawbar or just a length of chain attached to the implement. It was an inefficient arrangement, as the working depth could vary significantly where there was an undulating surface, field corners and headlands caused problems because reversing with the implement was either difficult or impossible, and the towing arrangement could also be dangerous.

The danger was due to the variation in the height of the hitch point on different tractors; when the hitch was too high, the forces from the implement could make

the front of the tractor rear up. This was long before the days of safety cabs, and many drivers were either killed or seriously injured because of tractors rearing over backwards. The Model F Fordson was one of the worst examples of this. In his book *Henry Ford and Grass-Roots America*, Reynold M. Wik estimates that 136 tractor drivers died and many more were injured on American farms in the five years ending 1922 due to Fordson tractors overturning.

IMPROVING IMPLEMENT LIFTS

Providing special hitch points to allow the implement to be attached directly to the rear of the tractor helped to solve some of the problems, and this was the arrangement chosen by a British manufacturer, Burnstead and Chandler

The first production Ferguson tractor was the Model A or Ferguson-Brown, built by the David Brown company, but designed and marketed by Harry Ferguson. It was the first production tractor equipped with the Ferguson System three-point linkage with hydraulic control.

of Birmingham, for its 24HP Ideal tractor. The Ideal was available for about nine years from 1912, and it had a number of novel features that helped to make it complicated and expensive. One of these was a manually controlled differential lock, allowing the operator to direct the power through either the left or right driving wheel only or both together, depending on soil conditions and traction requirements. Also, the driving wheels were equipped with spade

FERGUSON A

Manufacturer: David Brown Tractors
Model: Ferguson Model A
Production started: 1936
Power unit: Coventry Climax Type E, later replaced by David Brown engine
Power output: 20HP
Transmission: Gear drive with three forward gears and one reverse

lugs attached to a cam that retracted them automatically to prevent a build-up of mud in difficult soil conditions.

Implements were mounted on a frame attached to the rear of the Ideal, and this acted as a hinge point, allowing the implement to be raised above the ground for transport or during headland turns. The lift mechanism was a cable attached to the frame of the implement and powered by the tractor engine.

US manufacturers were also developing power-operated implement lifts. The Hackney brothers from St Paul, Minnesota, began building their Auto-Plow tractor in 1911, and it was even more unconventional than the Ideal. The design included two driving wheels at the front and a single wheel at the

rear for steering, and the high clearance frame allowed space for a mid-mounted plough beneath the tractor. The lift mechanism for the three-furrow plough was powered from the engine and controlled from the driver's seat. The Auto-Plow was one of the first two-way or bi-directional tractors, designed to work either forwards or backwards with two seating positions allowing the driver to face either way. Several versions were built, including one with a 40HP engine and a later model rated at 30HP, but production ended in about 1919. The Hackney Auto-Plow and the Nevada Auto Plow have so many similar features – including the three-wheel layout, the two-way operation, the mid-mounted plough with a power lift and look-alike styling – that it is difficult to believe they were not based on the same design. Production started in Nevada, Iowa, in 1913 and ended about two years later. Details of two models have survived, with 25HP and 60I IP engines.

Other American power lift pioneers

Waterloo Boy tractors were sold in Britain and Ireland under the Overtime name, and distribution in Ireland was controlled by Harry Ferguson's garage business. The Overtime in the photograph is owned by an English collector, but its first owner was a farmer in Ireland.

included the Lawter company from Newcastle, Indiana, and its 1914 model shared the same three-wheeler layout and two-way driving arrangement as the Hackney and Nevada models. The Lawter also used a mid-mounting arrangement for the plough, and the power lift mechanism was controlled by a foot pedal.

Deere & Co.'s contribution to the development of lift mechanisms started in 1927 with a mechanically operated system on the Model C, predecessor of the popular General Purpose or GP series. This was followed in 1932 by what was probably the first implement lift to be operated by hydraulic power on a tractor. It was available on the Model A rowcrop tractor, and it provided the operator with an easier method of controlling a mounted implement.

Harry Ferguson chose hydraulic power when he developed his system for lifting and controlling implements, and this was the equipment Henry Ford helped to introduce. Ferguson and Ford had much in common; they were both from farming families, but they both disliked the slow pace of working with animal power, and they were both eager to leave home to start a life away from the farm and develop their engineering abilities.

Ford left home to work in Detroit and Harry Ferguson went to Belfast in what is now Northern Ireland to work in his brother's cycle repair business. The Ferguson brothers expanded the business to include motorcycle and car repairs, helped by Harry Ferguson's aptitude for repairing the temperamental petrol engines of the early 1900s. Harry Ferguson later opened his own garage, and he also developed an interest in motor racing. In 1909, he became the first person in Ireland to build and fly his own aircraft.

THE FERGUSON SYSTEM

Ferguson's interest in tractors developed during World War I. Farmers in Britain were encouraged to use tractor power in order to increase food production and Ferguson, with a keen eye for a business opportunity, began selling tractors through his garage. The tractor he chose was the Waterloo Boy, known in Britain

as the Overtime, and he took a personal interest in the demonstrations and sales. His expertise with the tractors brought a plea for help from the Irish Board of Agriculture. Board members wanted him to visit the farms in Ireland where tractors were used, and to show the owners and the drivers how to make the most effective use of tractor power.

The tractor tour was completed during March 1917, and it was almost certainly this experience that encouraged Ferguson's interest in the way in which implements were attached to tractors. By September 1917, he had taken out a patent - the first of many - for an improved plough design. His next development was a plough designed specifically for the recently introduced Fordson Model F tractor, and this linked the plough to the tractor by a pair of struts designed to help maintain a more even working depth and also reduce the risk of the tractor overturning backwards. This plough was made in the United States, where it sold in substantial numbers.

Harry Ferguson's interest in implement attachment and control systems continued, and, during the late 1920s, he and his engineers developed a linkage with three attachment points and hydraulic operation. It was the basis for what was

Ferguson Model A production started with Type E power units supplied by Coventry Climax; by 1938, when this Model A engine was made, the David Brown company was building its own engine. The David Brown version of the four-cylinder engine featured a redesigned cylinder head.

to become known as the Ferguson System, still used as the standard implement attachment and control system on most of the world's farm tractors.

With the design of the attachment and control system virtually complete, Ferguson needed a special tractor he could use to demonstrate the benefits of the equipment in order to attract the interest of a manufacturer. The tractor he designed for the job was completed in 1933. It had a jet-black paint finish and, not surprisingly, was known as the Black Tractor. It was the first Ferguson System tractor and, at the time of writing, is on long-term loan to the Science Museum in London.

The gears for the Black Tractor were cut by David Brown, a company specializing in this type of work. It became increasingly interested in manufacturing a tractor equipped with the Ferguson System. This later led to a partnership, with Ferguson retaining sole responsibility for design and marketing, while David Brown formed a separate company to manufacture the tractor. The tractor produced by this partnership was the Ferguson Type A, often called the Ferguson-Brown. It was announced in 1936, and it inherited many of the design features of the Black Tractor.

Power for the first 500 Type As was provided by an 18-20HP Coventry Climax engine instead of the Hercules power unit used in the Black Tractor; a new David Brown 2010cc power unit was used for the rest of the production run. The black paint was replaced by battleship grey, the colour Harry Ferguson chose for all the production tractors with which he was closely associated. The styling of the two tractors was similar, however, and they both featured a three-speed gearbox and, of course, the Ferguson System linkage.

The Ferguson System with its power-operated implement lift system with automatic draught sensing to maintain a uniform ploughing or cultivating depth offered significant advantages. The Type A with a linkage-mounted implement easily outperformed more powerful tractors, but it was also more expensive than many of its competitors. Farmers also had to buy special Ferguson System implements to work with the Type A tractor, while other tractors including the much cheaper Fordson were designed to work with conventional machinery.

A MOVE ACROSS THE ATLANTIC

Sales of the new tractor were disappointing, and the lack of success put increasing strain on the Ferguson-Brown partnership. David Brown asked for design changes, including more power and a four-speed gearbox, to make the tractor more saleable, but Ferguson insisted the design was right and blamed quality control problems at the factory for poor sales performance.

Meanwhile, unsold tractors were piling up at the factory, and, as the situation continued to deteriorate, Harry Ferguson decided to visit the United States. The reason for his visit was to provide Henry Ford with a demonstration of the Type A tractor, but the background to the visit is still

shrouded in mystery. It has been suggested that Ferguson asked for the opportunity to show his new tractor to Ford, but it is also possible that Ford had heard about the revolutionary new implement attachment and control system and wanted to see it in action. What we do know is that Ferguson did not tell his business partner, David Brown, the reason for his sudden departure to the United States.

Henry Ford may have regretted his decision to transfer production of the Model F tractor from the United States to Ireland in 1928, and subsequently to England. The updated Model N imported from England was still selling in small numbers in the United States, but it no longer dominated the market as the Model F had done, and, by the mid-1930s, Ford was keen to re-establish his position as a major force in the tractor industry. Several experimental prototype models were produced, and Henry Ford took a keen interest in the

progress of the new tractor project – some of the surviving photographs of the time show him at the controls.

One of the experimental tractors was powered by a V-8 engine designed for a Ford truck, and this machine was demonstrated to a group of journalists from farming magazines. Ford used the occasion to announce his interest in getting back into tractor production as soon as possible; 'I don't care if we can't make a cent of profit,' he said. 'The main thing is to get something started.'

It was obviously an appropriate time for Harry Ferguson to arrive on the scene with his tractor and implements. The demonstration was held in a field close to Fairlane, the Ford family mansion on the banks of the River Rouge near Dearborn. An Allis-Chalmers Model B tractor and a Fordson Model N were brought from the Ford farm to provide a comparison, and, as usual, the Ferguson System tractor won an easy victory.

The basis of the Ferguson System of implement control is the hydraulically operated three-point linkage with draft control, shown here on a British-built TE series Ferguson tractor. It was one of the major milestones in farm tractor development.

FORD 9N

Manufacturer: Ford Motor Co.
Model: 9N
Production started: 1939
Power unit: Ford four-cylinder L-head with 1966cc (120 cu. in) displacement
Power output: 23HP at 2000rpm
Transmission: Gear drive with three forward gears and one reverse

It was the famous handshake agreement between Henry Ford and Harry Ferguson that resulted in the Ford 9N tractor, available from 1939. The combination of Henry Ford's manufacturing resources and the unique Ferguson System made the 9N one of the industry's biggest successes.

Henry Ford was impressed by the performance of the Ferguson System tractor, and he was probably equally impressed by its inventor. As well as possessing similar backgrounds, both men shared the belief that efficient, low-cost tractor power had an important part to play in helping to restore agricultural prosperity. They both genuinely wanted to help the agricultural industry, which was facing serious financial problems that were forcing farmers on both sides of the Atlantic out of business.

The meeting between Ford and Ferguson ended with an extraordinary agreement that committed both men to work together to produce a new Ferguson System tractor. This would involve a massive financial investment

by Henry Ford, while Harry Ferguson contributed the patents which represented almost 20 years of work. The agreement was simply sealed with a handshake, and the precise details were never witnessed or recorded, and there was no formal contract. Under the agreement, which appears to have been suggested by Ferguson, Ford would use his financial and technical resources to build the tractors, while Ferguson would be responsible for the design and the marketing. This was a similar arrangement to the Ferguson-David Brown partnership that was already in the process of falling apart, and it is perhaps surprising that Ferguson should have modelled his new agreement on one that was obviously flawed.

A NEW PARTNERSHIP

In fact, the new partnership worked well, initially. It produced the brilliantly successful 9N, 2N and 8N tractor series, it established the Ferguson System as the industry standard for implement attachment and control, and there were important benefits for the farming industry. It was only after the ageing Henry Ford handed over the control of his vast company to others that the

friction started, ending in the biggest and most expensive legal battle the tractor industry has ever known.

The immediate result of the 'handshake agreement', as it is called, was that Henry Ford set up an engineering team to start developing the new tractor, and its first priority was to

Part of the settlement in the Ferguson/Ford lawsuit was an agreement by the Ford Motor Company to modify the rear linkage and hydraulic system on the new 8N tractor to avoid any further patent infringement. In spite of this, the US-built 8N was immensely popular.

strip down the Ferguson Type A to see how the hydraulic system and the three-point linkage operated. Meanwhile, Ferguson returned to Britain for a brief visit to sort out his affairs in preparation for a prolonged stay in the United States.

The affairs needing his attention included bringing an end to his partnership with David Brown. This may have been welcome news for David Brown, who had already started the design work for the improvements he wanted for the Ferguson A. Instead, the extra power, the four-speed gearbox and an innovative seat for two people featured in a new tractor designed, built and marketed by David Brown Tractors. Just over 1200 of the Type A tractors had been built, and some of these were still in stock when the new David Brown model arrived in 1939. It was called the VAK1A, and it was the start of a success story that put David Brown among the world's leading tractor companies. It was eventually taken over by the US

Above: When the 9N tractor was replaced by the improved 8N model, it marked the end of the old handshake agreement. The Ferguson System hydraulics and rear linkage included in the specification, without Harry Ferguson's consent, resulted in a famous lawsuit.

company Tenneco and merged with its Case tractor business.

When Ferguson returned to Dearborn, the development work on what was to become the 9N model was already progressing with extraordinary speed. Production versions of the new tractor were available in June 1939, only eight months after the handshake agreement that had started the project. On 29 June, almost 500 guests watched the new tractor in action at a reception and demonstration at the Ford farm at Dearborn. The 9N, or the Ford Tractor with Ferguson System to use its correct name, was a big success. It was painted battleship grey, presumably at Ferguson's insistence, and it was equipped with a three-speed gearbox. The four-cylinder engine with a side-valve layout was based on half a Ford V-8 engine from a Mercury car, an arrangement that avoided the cost and time needed to develop and test a new engine for the tractor. It also helped to reduce the production cost. The 1966cc power unit developed 23HP and was equipped with a battery, a generator and an electric starter motor, and rubber tyres were also included in the standard specification. The 1939 price in the United States was $585, compared with $495 for the most basic version of the English-built Model N Fordson. Production totalled 10,000 tractors by the end of the first year, and almost 43,000 Model Ns came off the assembly line in 1941. During the following year, however, there were production delays due to wartime shortages of components and raw materials.

To keep the production line going, a reduced specification version called the 2N was introduced in 1942 with steel wheels instead of rubber tyres and with a magneto to replace the battery and electric equipment. Another version was the 9NAN, modified to burn paraffin instead of petrol and aimed mainly at the British market. The peak year for the 9N tractor was 1946, its last full production year, when 74,004 tractors were assembled.

Henry Ford had helped to set up Harry Ferguson's distribution company by making a substantial loan to cover some of the initial costs, and channelling both the tractors and the Ferguson System implements through the company proved a profitable business. Building the tractors was less profitable,

or so the Ford company claimed, but this was not a threat to Ferguson's business as long as Henry Ford was in total control of his company.

AN ACRIMONIOUS SPLIT

In 1945, when he was 82 years old and no longer capable of running one of the world's great industrial concerns, Henry Ford handed over the reins to his grandson, Henry Ford II. By this time, the Ford company was losing substantial amounts of money, partly because of problems resulting from wartime disruption and shortages, and partly because of the lack of firm leadership when the ageing Henry Ford was in control. Urgent action was needed to cure the company's financial problems, and the tractor operation was one of the

ALLIS-CHALMERS U

Manufacturer: Allis-Chalmers
Model: Model U
Production started: 1929
Power unit: Continental, later replaced by A-C four-cylinder engine
Power output: 34HP (A-C engine)
Transmission: Gear drive with three forward gears and one reverse

areas that was leaking money. Not surprisingly, the new management team decided the Ford company should benefit from the profits available from distributing the tractors, and it gave Harry Ferguson six months' notice of its intention to stop supplying tractors to his company. This left Ferguson with a big and efficient distribution organization in the United States, but no tractors to sell. From Ferguson's point of view, worse was to come when Ford decided to stop making the 9N tractor and replace it with a new model.

The new tractor arrived in July 1947, just after the distribution agreement came to an end. It was called the Ford 8N and the specification included the full Ferguson System linkage and hydraulics. The 8N also included 20

design improvements, according to Ford publicity, and the most important of these was a new four-speed gearbox, while the most visible alteration was the addition of some red paint to help brighten up the grey finish chosen by Ferguson.

Harry Ferguson responded vigorously to the Ford company's actions. He set up his own factory in the United States to supply Ferguson tractors to his distribution company. He was able to supply sufficient Ferguson TE-20 series tractors from the recently established British production line at the Banner Lane factory near Coventry, England, to keep

the distribution business operating until his American-built tractors were available. His other response was a lawsuit against the Ford company, with claims amounting eventually to about $340 million, an enormous sum by 1940s standards. It included compensation for damage to his distribution company after the supply of Ford tractors ended, and he also claimed for patent infringement through the unauthorised use of the full Ferguson System on the 8N tractor. The legal battle which followed was complicated, acrimonious and immensely expensive, involving up to 200 lawyers at one stage and an estimated one million documents.

Some of the difficulties were caused by the nature of the original handshake agreement, making it difficult to prove

The Allis-Chalmers Model U established its place in farm mechanization history as the first production tractor to be equipped with low-pressure inflatable rubber tyres. The tread designed for tractor tyres in the early 1930s, shown in the cutaway picture of a Model U, is similar to modern turf tyres.

exactly what had been decided and the terms on which the deal could be terminated. Henry Ford had died in 1947 at the age of 83, and this left Harry Ferguson as the only person who knew what was covered in the agreement. The lawsuit was filed in 1948, but settlement was not reached until April 1952, by which time Ferguson's distribution business was operating profitably once more due to the success of the British-built TE-20 tractors and the American-built TO-20

versions, causing the court to dismiss his claim for compensation. He finally accepted a $9.25 million settlement to cover the use of his patents on the 8N tractor, and the Ford company was also ordered to modify the 8N hydraulics to avoid further patent infringement.

A NEW GENERATION

Meanwhile, the new-generation Ferguson System tractors were increasingly popular. Production of the 8N peaked at more than 100,000 per

The Benz company, famous now as part of Mercedes-Benz, was one of the pioneers of diesel power; in 1923, its two-cylinder engines were used to power the Benz Sendling, the first production tractor with diesel power. More diesel-engined Benz and Mercedes-Benz tractors soon followed.

year in both 1948 and 1949, and the combined total for the TE-20 and TO-20 Ferguson tractors – both closely based on the original Ford 9N design – also briefly topped 100,000 per year. Only one other tractor has broken through the 100,000 per year production barrier, and that was Henry Ford's Model F tractor at the peak of its popularity during the 1920s.

The Allis-Chalmers Model U tractor failed to come anywhere near the Ford/Ferguson production figures. It did, however, introduce one of the most important design improvements in the 1920s and 1930s, and in the process it featured in one of the most unusual and entertaining publicity campaigns the tractor industry has ever known.

Francesco Cassani was the first Italian to build a diesel tractor. His Cassani tractors were produced in small numbers during the late 1920s, but attempts to obtain the financial backing he needed for volume production failed and he concentrated on engine design and development instead.

When the Model U arrived on the market in 1929, it was known as the United tractor, named after the United Tractor and Equipment Co. of Chicago, an association formed by some of the medium-sized companies making farm equipment. They had joined forces to challenge the big full-line manufacturers such as International Harvester and Deere. Allis-Chalmers was a member of the group, and its contribution to the United product range was building a new mid-range tractor powered by a 35HP Continental petrol engine.

THE SWITCH TO RUBBER TYRES

Although the United company failed to survive, the Allis-Chalmers tractor business was more successful, and the tractor remained in production as the Model U, powered by a 34HP Allis-Chalmers petrol/paraffin engine instead of the original Continental unit. It was originally a conventional late 1920s design with steel wheels equipped with spade lugs, but it became the first tractor in the world to replace the steel wheels with a new type of low-pressure tyre designed specifically for tractors.

Steel wheels equipped with various types of lugs or cleats for increased grip in soft ground conditions were standard equipment for the first 40 years of tractor history. Rubber tyres had been tried as an alternative, but these were high-pressure truck-type tyres that were suitable for travelling on the road, but lacked the flexibility to grip the soil efficiently for field work. There were also a few tractors equipped with solid rubber tyres, but these were mainly industrial tractors used primarily on roads and other hard surfaces.

One of the disadvantages of steel wheels was the way in which they transmitted all the bumps and vibrations from the road to the rest of the tractor and the driver. Not only did this make them uncomfortable, but also excessive vibration could damage the mechanism of the tractor, shaking bits off and breaking castings. Hence the maximum travel speed for tractors on steel wheels was often as low as 4km/h (2.5mph) and rarely more than 8km/h (5mph). Another problem was

that the lugs or cleats, essential for field work, had to be removed or covered with smooth steel bands to avoid damage when travelling on the road. Fitting or removing either the lugs or the road bands was an unpleasant and time-consuming job, especially when the wheels were caked with mud. The only alternative on tractors used for general farm work was to use crawler tracks, but, as these were also made of steel, they shared some of the disadvantages of steel wheels and, in

The Caterpillar company made an important contribution to establishing the benefits of diesel power on American farms. Caterpillar used a small petrol engine as the starter motor for the main diesel engine; this picture shows the petrol engine on a 1939 Caterpillar D2 tractor exported to Britain.

addition, lacked the flexibility of being able to work with road bands.

Attempts to solve the steel wheel problem met with little or no success until 1932, when tests showed that rubber tyres designed for low inflation pressures would absorb much of the vibration when travelling on the road and were flexible enough to grip the soil for jobs such as ploughing. The company that made the breakthrough was Allis-Chalmers, and its first experiments on a Model U tractor are said to have involved tyres borrowed from a freight aircraft. The leading tyre manufacturers were quick to welcome the prospect of a new market for their products, and a Model U with an experimental set of special tractor tyres was supplied to a dairy farm near Waukesha, Wisconsin, for evaluation. The farmer was enthusiastic, noting

particularly that the tyres did not cause the damage to the surface of his grass fields he had experienced with steel lugs.

Model U tractors equipped with the new tyres were available during the summer of 1932, and the first customer was a farmer from Dodge City, Kansas, who bought a new model U on rubber tyres after seeing a demonstration. Unfortunately, the Dodge City farmer's enthusiasm was not widely shared, and demand for the new tyres was disappointing. There were worries about the

CATERPILLAR D2

Manufacturer: Caterpillar
Model: D2
Production started: 1938
Power unit: Cat four-cylinder diesel with 95.3 x 127mm (3.75 x 5.0in) bore and stroke
Power output: 29HP rated
Transmission: Gear drive with five forward gears and one reverse

Model U tractor, and almost 50 per cent of all the new tractors sold in the United States were equipped with rubber tyres by 1937.

Tractor tyre development in Britain also started in 1932, when the Dunlop Rubber Co. began experimenting with inflatable tyres. Production versions were available during the following year, and the company won a silver medal from the Royal Agricultural Society at the 1934 Royal Show, in recognition of the importance of the new tyres. French tractor tyre development began in 1933, when Michelin instituted a research programme, with

risk of punctures; doubts about the wear rate and replacement cost of rubber tyres used in the field persuaded some farmers to specify steel wheels. A publicity campaign was needed to promote the benefits of the new rubber tyres.

Britain's first commercially successful diesel tractor was the Field Marshall, powered by a slow-revving, single-cylinder engine with a distinctive exhaust note. Later versions, such as the 1949 Series 3 tractor in the photograph, were equipped with a cartridge starting system.

WINNING OVER THE BUYING PUBLIC

Allis-Chalmers decided to focus their advertising on speed, as the smoother ride offered by tractors on rubber tyres allowed faster travel speeds on the road. Top speed of the standard Model U with a three-speed gearbox and steel wheels was 5.5km/h (3.5mph), but an extra fourth gear was available for tractors fitted with rubber tyres. This boosted the speed to 24km/h (15mph), and this was featured in advertisements with the slogan '5 miles an hour on the plow – 15 miles an hour on the road!'.

Something extra was still needed to give the tyres a more positive image, and the Allis-Chalmers engineers were given the job of modifying one of the tractors to achieve a much faster top speed. The new high-speed tractor made its first appearance at the Milwaukee Fair in 1933, where it was demonstrated with a plough before being driven on to the fairground racetrack to complete a lap at an average speed of 57km/h (35.4mph). The speed was officially

timed and recorded, and it was accepted as a new world record for tractors.

More hot-rod Model Us were prepared, and these formed a racing team with some of the United States' most famous car racing drivers at the wheel. The great names from the racetrack included Lou Meyer, Ab Jenkins and Barney Oldfield, and, as the racing team toured the state fairs and other big events during the summer of 1933, they attracted an estimated one million spectators and large amounts of press coverage. The engineers coaxed more speed out of the tractors, and the publicity campaign ended when Ab Jenkins drove a Model U at 108km/h (67mph) on the Utah salt flats. This performance was accepted as yet another new world record for a farm tractor, breaking the previous 103.45km/h (64.28mph) record set in Dallas, Texas, a few weeks earlier by Barney Oldfield, at the wheel of a Model U.

The publicity campaign was a success. There was a sharp upturn in sales of the

the production version available in the following year. Kleber, now a Michelin subsidiary, announced the first radial tyres for tractor driving wheels in 1961.

DIESEL POWER

Another important technical development in the 30-year period from about 1920 was the development of diesel-powered tractors. This is another example of a development that has made a major contribution to tractor performance and is featured on virtually every modern tractor.

Diesel tractor development began in Europe, and the first company to offer a

tractor with a diesel engine was Benz of Mannheim, in Germany. Benz began experimenting with diesel power after World War I had ended, and small-scale production started in 1921 or 1922. The engine was a twin-cylinder design, used initially in a small truck; in 1923, the same type of engine was offered as an option for the Benz-Sendling motor plough. This was a strange contraption with a single driving wheel at the rear. It had been available since 1919 with a petrol engine, but the 1923 version had the distinction of almost certainly being the world's first diesel-powered tractor. The Benz range of diesel engines also

included a single-cylinder version with hopper cooling, and a four-cylinder diesel arrived in 1923 to be used as a truck engine. All the early Benz engines were designed with a pre-combustion chamber, a feature first patented in 1909 and developed to improve combustion efficiency. In 1926, the Daimler and

The ill-fated Agricultural and General Engineers company produced one of the most advanced diesel tractors available in the early 1930s. The Aveling-powered tractor pictured took part in the 1930 World Tractor Trials and also set a new world record by ploughing for 977 hours non-stop.

Benz companies, named after the two earliest pioneers of the car industry, merged to form Daimler-Benz, now known as Daimler-Chrysler. The Benz name survives in Mercedes-Benz, one of the world's most prestigious marques.

Benz continued to make diesel-powered tractors for about five years after the merge; these were exported to a long list of countries, including Australia and Britain. The Mercedes-Benz name returned to the tractor market in the late 1940s with the Unimog all-terrain tractor/transport vehicle, followed in 1972 by the innovative MB-trac tractor range.

AN ITALIAN DIESEL PIONEER

The next diesel tractor pioneer was Francesco Cassani of Italy. His engineering skills were acquired in his father's workshop in Treviglio, near Milan, while helping to make and repair steam engines and other machinery for local farmers. His special interest was diesel power, and, in 1926, when he was still only 19 years old, he had built his own two-cylinder diesel engine. In the following year, it was demonstrated at a local agricultural college in a tractor he had also designed and built. Cassani made several attempts to find a manufacturer to build the Model 45 tractor commercially, but only about 15 were built.

Instead of developing more diesel tractors, Cassani concentrated on

INCREASING HORSEPOWER

Farmers and contractors are demanding more tractor power to help increase work rates and improve efficiency, but increasing engine power is a relatively recent development and there was little change during the first 60 years or so of tractor history.

The first generation of American tractors in the period before 1910 averaged about 20 to 30HP, while the British average was an estimated 15 to 20HP, and there was little change during the mid 1920s when the 20HP Fordson Model F dominated the tractor market. By about 1950 the Ford 8N in America and the TE-20 series Ferguson in Britain were big sellers, and the average engine power for new tractors in America had probably passed the 40HP mark and had reached about 30HP in Britain.

The big power increases started in the 1960s. The average engine power for new tractors sold in Britain had reached 63HP in 1975, with a further boost to 87HP by 1990, reaching about 116HP in 2000.

designing diesel engines for other purposes, and his successes included marine engines and even a rotary diesel engine to power aircraft. The big increase in farm mechanization after the war ended in 1945 encouraged Cassani to design a new tractor. He formed a new company to build tractors under the Same brand name, and the first product was a lightweight three-wheeler powered by a 10HP single-cylinder petrol engine.

When the first diesel-powered tractor arrived in about 1951, it was powered by an air-cooled engine, and Same remained one of the leading manufacturers of this type of power unit. Air cooling is less suitable for big diesel

John Deere's first production tractor with diesel power was the Model R. The familiar two-cylinder horizontal engine layout was not abandoned in the switch to diesel power, and boosting the number of forward speeds from two to five increased the complexity of the transmission.

engines, and the trend towards more powerful tractors in the 1990s encouraged the switch to liquid cooling. The company, now known as Same Deutz-Fahr claims to be number four in world tractor production and builds tractors in Italy, Germany and Poland under the Deutz-Fahr, Hurlimann, Lamborghini and Same brand names.

DIESEL IN NORTH AMERICA

Caterpillar was the first US tractor manufacturer to offer diesel power. The first experimental diesel engines were converted petrol units, but the field testing with a specially designed Caterpillar diesel engine started in 1930 with a diesel version of the Sixty model.

The production version of the Diesel Sixty arrived in 1931, and it set a new fuel efficiency record when consumption was measured independently in the brake horsepower test at Nebraska. The efficiency figure was 3.05 horsepower hours per litre (13.87 horsepower hours per gallon) of fuel, and the four-cylinder I-head Caterpillar engine had a 700rpm rated speed and produced a maximum of 77.08HP.

The torque characteristics of diesel engines are particularly suitable for heavy-duty pulling jobs that are often given to crawler tractors, and this factor plus the reduced fuel consumption of the new diesel engine attracted customers to the Diesel Sixty. More diesel models

were added to the Caterpillar range during the 1930s, starting with the Diesel Thirty-Five, Fifty and Seventy-Five models announced in 1932. Some of the diesel tractors were identified with a model number starting with the letters RD, chosen, it is said, as a tribute to Rudolf Diesel, who invented the engines that bear his name.

By 1937, most models in Caterpillar's range were available in diesel-powered versions, and this includes the small R2 and D2 models. The R2's engine was fuelled by petrol or paraffin, and the D2 model was a diesel; however, both engine types shared the same cylinder block, the same 95.25 x 127mm (3.75 x 5in) bore and stroke, and they all produced 29HP rated output. The R2 and D2 models were tested at Nebraska in 1939; fuel efficiency figures recorded for each of the three fuels provide a useful comparison. In the rated output belt test, the R2 tractor burning petrol produced 1.93 horsepower hours per litre (8.76 horsepower hours per gallon) for the rated load belt test, the figure for the paraffin version was 1.83 (8.33) and the diesel-powered D2 was well ahead with 2.93 (13.32).

Caterpillar, more than any other company, established the success of

diesel power for farm tractors, and the company deserves great credit for recognizing the potential advantages of diesel power for tractors and for producing such a successful series of engines. During the 1930s and for almost 20 years following the launch of the Diesel 60, Caterpillar was easily the world's biggest and most successful manufacturer of diesel tractors.

The success of the Cat diesels forced other crawler tractor manufacturers in the United States to follow their example. The TracTracTor T-40 was the first model in International Harvester's McCormick-Deering range with a diesel engine. The four-cylinder I-head engine with 120.7mm (4.75in) bore and 165.1mm (6.5in) stroke developed its rated power at 1100rpm, and it set a new record for fuel efficiency when it

As diesel power's popularity increased from the late 1940s onwards, the distinctive Perkins badge became increasingly familiar on a long list of tractors. The badge in the photograph is on a Farmall BMD tractor built in 1951 at the International Harvester factory in Doncaster, Yorkshire, UK.

achieved 15.18 horsepower hours per gallon in the belt test at Nebraska in 1935. International Harvester followed this with the WD-40 diesel model, probably the first US-wheeled tractor with diesel power. It was available from 1934, producing slightly less than 50HP.

Cleveland Tractor Co.'s first diesel tracklayer in their Cletrac range was the 40 Diesel, announced in 1934 and powered by a Hercules I-head engine with six cylinders. The rated output was 57HP. The first diesel model from Allis-

Chalmers was the WK-O crawler tractor, tested at Nebraska in 1937.

The D2, like other diesel models in the Caterpillar range during the 1930s, used a small petrol engine to start the diesel. Using different fuels for starting and running the engine was obviously a complication – although it was one that drivers of tractors with petrol/paraffin engines were familiar with – and the small petrol or donkey engine made the diesel tractor more expensive. In spite of the extra cost, the donkey engine was the starting method chosen by most US manufacturers until well into the 1950s.

It was the system chosen by John Deere when it introduced its first diesel, the Model R, announced in 1948 and produced for six years from 1949. Deere engineers had done some preliminary development work on a diesel engine as early as 1935, but design work on the Model R engine began in 1944. For the production version, the design team remained loyal to the two-cylinder horizontal engine layout that had powered virtually all previous John Deere tractors. The new diesel was a big engine with 146 x 203mm (5.75 x 8in) bore and stroke, and maximum power output was 51HP, making the Model R the most powerful production tractor John Deere had built. It was also the most economical, setting yet another record for fuel efficiency at Nebraska.

The petrol engine for starting the big diesel was also a two-cylinder design, but in this case the cylinders were horizontally opposed, rather than parallel. The petrol engine was water-cooled and shared the same cooling system as the big diesel, allowing the heat produced by the petrol engine to warm up the cooling system and make the diesel engine easier to start on cold mornings. An electric starter motor was provided for the petrol engine, powered by a battery, and another starting system – this time manually operated by lever – was included as a back-up if the battery was flat. These thorough but complicated starting arrangements presumably added to the Model R's price, but it was popular in spite of this.

Hand cranking is not an option for starting a big diesel engine – the high compression makes the engine too difficult to turn over. Electric starting with a motor powered by a battery was also unsatisfactory as early diesels were not easy to get going and batteries available at that time lacked the capacity to crank a temperamental diesel on a cold morning. The cold start problem for diesel tractors would eventually be solved by improvements in the combustion chamber design, when electric heating elements were added to warm the cylinders and easily combustible gases were injected into the chamber, but these developments came later.

There are three power units in the engine compartment of the John Deere Model R. The main engine is the twin-cylinder diesel unit developing 51HP. A small petrol engine with two horizontally opposed cylinders is the starter motor for the big diesel, and there is also a battery-powered electric motor to crank the petrol engine.

Another option available in the United States, and attracting some interest in the 1930s, was the diesel engine with fuel injection plus an electric system and spark plugs for starting on petrol. The Waukesha-Hesselman diesel engine that made a brief appearance in the 1930s on tractors made by the Bates Machine and Tractor Co. of Joliet, Illinois, used this arrangement. The 35 and 40 models both featured this engine design, but the numbers produced were small.

International Harvester used a similar engine arrangement on their TracTracTor T-40 model and the WD-40 wheeled version available from 1934. The engine was made by International Harvester, and they also made a slightly smaller version with the stroke reduced by 6.35mm (0.25in) for the TracTracTor TD-35 announced in the following year.

BRITAIN'S ENTRY INTO DIESEL

British interest in diesel power for tractors started in about 1928. That was when the Marshall company began its development work on what later became Britain's first commercially successful diesel engine when it was used to power the Field Marshall series tractors. The engine was a true diesel, but it was based on the Lanz semi-diesel and inherited the Lanz engine's single-cylinder design. In its early versions, the Marshall engine's single-cylinder bore and stroke were 203mm (8in) and 267mm (10.5in), the rated speed was a modest 550rpm and the maximum power when the tractor was tested independently was 29.1HP, comfortably above the 24HP rated figure.

The engine was started by using a blow lamp, but an improved version with cartridge starting was available from 1945 when the Series 1 tractor was introduced. Production continued until 1957, and included the VF series crawler version. Single-cylinder Marshalls became popular, and the distinctive exhaust note of the big, slow-revving engine was a familiar sound on many British farms.

The Marshall company, based at Gainsborough, Lincolnshire, in England had been one of the most successful of

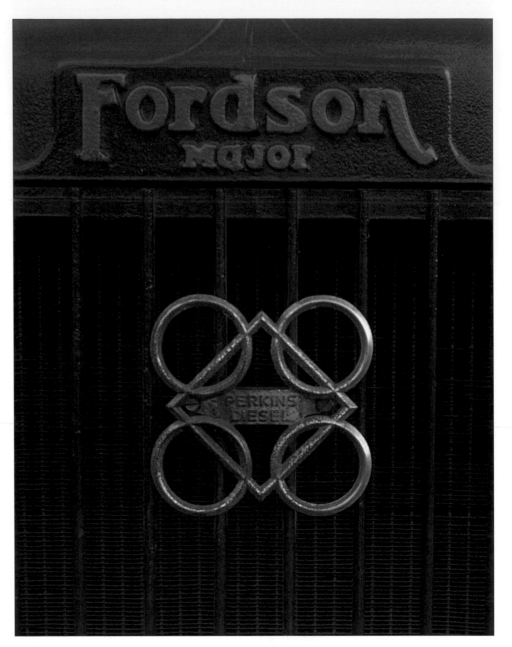

Britain's many agricultural steam engine manufacturers. J. & H. McLaren of Leeds was another company with a steam engine background. By the late 1920s, McLaren was firmly committed to diesel power, importing Mercedes-Benz tractors from Germany and also assembling the twin-cylinder version for sale under the McLaren name.

Maximum output from the two-cylinder vertical engine in the McLaren tractor was 30.5HP, and the rated engine speed was 800rpm. The bore and stroke figures were 134.6mm (5.3in) and 200.7mm (7.9in), and the engine was started by using a blowlamp. Blowlamp starting was also required with the single-cylinder version in the Mercedes-Benz model. The dimensions of the

This is another example of the spread of Perkins power. This time the badge is on the front of the Fordson E27N Major, the model that replaced the ageing Model N at Ford's British factory in Dagenham, Essex. Perkins engines were used for Fordson Major customers who preferred diesel.

horizontal cylinder were 149.9 x 240mm (5.9 x 9.45in) and the cooling system was based on a hopper.

Another British company with faith in the future of diesel-powered tractors was Agricultural and General Engineers (AGE). It was formed by a large number of farm equipment manufacturers joining forces to provide a British-built full line of tractors – and machinery to compete with the big US and Canadian

When the Turner Manufacturing Co. developed their new Yeoman of England tractor, it used a four-cylinder diesel engine built specially for the tractor and designed by a leading diesel engine specialist. The cylinders are arranged in a V-formation with a 68-degree angle.

companies. In commercial terms, AGE was a disaster, but one of its few genuine achievements was the development of two of the most advanced diesel tractors in the world.

One of the AGE tractors was powered by a four-cylinder Blackstone engine. The rated output claimed by the manufacturer was 26HP, but the maximum output when the tractor was independently tested was 37.7HP, suggesting that Blackstone had seriously underestimated its performance. The rated speed of the engine was 1000rpm, and a petrol engine was provided for starting the diesel. The Blackstone engine powered a tractor made by Garrett, another AGE member with a prestigious history based on steam engine power.

The other diesel engine from the doomed consortium was made by Aveling and Porter of Rochester, Kent, yet another of the former steam engine manufacturers. It was also a four-cylinder design, but this time the rated and actual power output figures were closer together at 38HP and 42.4HP, respectively. The engine, called the Invicta, had an 1150rpm rated speed, making it one of the fastest of the early generation of tractor diesels. The starting equipment was an electric motor, suggesting the designers had solved at least some of the traditional diesel starting difficulties.

The Marshall diesel, the Mercedes and McLaren models and the two AGE tractors all took part in the 1930 World Agricultural Tractor Trials held near Wallingford, Oxfordshire. This was one of the biggest events of its kind ever held, organized jointly by the Oxford University's Institute of Agricultural Engineering and the Royal Agricultural Society of England. The trials attracted 33 tractors and three small powered cultivators entered by manufacturers in eight countries, with the United States leading the league table with 12 tractors entered. The test programme took 24 hours for each tractor, and the trials started on 2 June and were eventually completed on 26 July.

THE 1930S WORLD TRIALS

For tractor historians, the 1930 trials are a mine of information, providing a mass of carefully measured performance data. A special feature of the trials is that they included five of the earliest diesel tractors to provide a comparison with a large number of petrol and paraffin models, plus a small number of semi-diesels. The fuel efficiency comparisons are particularly interesting, but the figures, based on imperial gallons, should not be compared with data from the tests at Nebraska, where US gallons were used for the fuel consumption calculations.

Ten of the tractors completing the full test programme were petrol powered, and these returned the poorest fuel efficiency figures in the rated load belt test, with 2.08 horsepower hours per litre (9.46 horsepower hours per gallon). The best figure in this section was 2.29 (10.40) from the British-designed but French-built Austin. In the paraffin section, there were nine completed test reports, with the fuel efficiency figures averaging 2.16 horsepower hours per litre (9.84 horsepower hours per gallon). They ranged from (1.29) 5.86 – the worst performance in the trial – for a Rushton tractor, which was a British copy of a Fordson Model F, to a highest figure of 2.47 (11.2). Three US tractors a Massey-Harris 12/20, Case Model C and an International 15/30 – tied for top place by scoring 2.47 (11.23).

Four semi-diesels averaged 2.76 horsepower hours per litre (12.55 horsepower hours per gallon), with an HSCS or Hofherr-Schrantz-Clayton-Shuttleworth from Hungary easily topping the fuel efficiency table with 2.97 (13.55). Easily the best figures came from the five tractors in the diesel section, where the average score was 3.59 horsepower hours per litre (16.31 horsepower hours per gallon), and the single-cylinder Marshall achieved a remarkable 3.80 (17.30), the best result in the trials.

In spite of producing such a convincing record for fuel economy, the diesel tractors failed to make much impact on the market. Of the five tractors competing in this section of the trials, only the Marshall was still available three years later. This was partly due to the collapse of the AGE group, which took some of the member companies with it. One reason for the apparent lack of interest in diesels was price. The list price of the AGE's Aveling & Porter–powered tractor with 38HP rated output was £525 according to the 1930 trials catalogue, but customers were also offered the 36HP International 15/30 rated at 36HP with a petrol/paraffin engine for £320. A 40HP Case Model L, the most powerful wheeled tractor in the trials, was available for £348. McLaren was expecting customers to pay £360 for the hopper-cooled 20HP Mercedes-Benz diesel, but the International Harvester 10/20 and Massey-Harris 12/20, both with 20HP rated output, were both in the price list at £220. The 20HP Austin was offered at £210; this was undercut by the £209.50 price for its rival, the 20HP Rushton. Best value among the diesels in pounds per horsepower terms was the single-cylinder Marshall, listed in the trials catalogue at £315 and 24HP, while AGE's 26HP Blackstone-powered diesel and the Benz-powered McLaren with 27.5HP available were both listed at £500.

Aveling & Porter survived the AGE disaster, but the economy measures introduced to help the company's recovery programme included scrapping the Invicta diesel engine development programme. Charles Chapman's responsibilities as head of engineering at

This BR tractor was built in 1937, the year before a leading US industrial designer produced a new 'styled' look for the John Deere tractor range. The new styling with more rounded lines arrived in 1938, with the A and B models being the first to benefit from the new look.

Aveling & Porter included overseeing the Invicta programme, and he was aware of the engine's potential. In 1932, when the development programme ended, he and another senior Aveling & Porter executive named Frank Perkins left the company to start their own diesel engine business at Peterborough, England. The company was called F. Perkins Ltd, and it played a leading role in developing improved diesel technology, including higher engine speeds and easier starting characteristics.

A DEVELOPING DIESEL MARKET

During the 1930s, the biggest market for diesel engines was in trucks and buses, and the demand for multi-cylinder diesel engines for tractors did not develop until the postwar period. The Perkins P6 engine played an important part in developing this sector of the market. It was the first six-cylinder engine Perkins had built, available since 1937 and equipped with reliable electric starting, and when tractor makers were looking for a diesel engine after World War II, the P6 was the ideal answer. Ford offered the P6 as a more powerful option for its new Fordson E27N Major tractor. In 1947, Massey-Harris looked at all the available diesels before specifying the P6 for the MH 744PD tractor, the British-assembled and diesel-powered version of the MH 44 model.

The P6 became Europe's most popular diesel engine for tractors, and manufacturers such as Allis-Chalmers

and Ferguson also specified the smaller P series diesels, including the three-cylinder P3. Other Perkins customers included Bristol, Nuffield, Howard and the International Harvester factory in Britain. Perkins was taken over by Massey Ferguson in 1959, but is now owned by Caterpillar. At the time of writing, the list of tractor manufacturers using Perkins power includes Case IH, JCB, Landini and Massey Ferguson.

Perkins did not have the postwar tractor engine market to itself for long. David Brown, the company Ferguson had abandoned so abruptly in 1938, was one of the first tractor manufacturers to predict the growing

importance of diesel power. Its first four-cylinder diesel engine was available in 1949 as an option for the Cropmaster tractor. It was a genuine high-speed diesel unit, delivering its 34HP rated output at 1800rpm, and electric starting was standard.

In 1951, Britain's leadership in the development of high-speed, easy-starting diesel power for tractors, already established by Perkins and David Brown, was reinforced by the launch of the new Ford engine. It was the result of a development programme that had started in about 1944, and it first appeared at the 1951 Smithfield Show as an option in the new Fordson

E1ADKN Major tractor. The Ford petrol, paraffin and diesel power units were all based on a four-cylinder design with overhead valves. The capacity of the diesel version was 3610cc, the compression ratio was 16:1 and the specification included electric starting.

There were, of course, some less successful ventures into diesel power. The award for the most interesting of the British tractor diesels in the postwar period would go to the Turner Manufacturing Co. of Wolverhampton, England, for the engine in their Yeoman of England tractor. The engine was designed by Freeman Sanders, who was previously responsible for the

Massey-Harris replaced its previous grey paint colour with an eye-catching shade of red, and the old angular lines made way for a more streamlined shape. One of the first tractors to benefit from the switch to a brighter image was the new 101 Junior model built between 1939 and 1946.

development of Fowler engines, and he chose a V-4 layout for the new Turner engine. The cylinders were arranged at a 68-degree angle, and this left the cylinder heads bulging out of the engine compartment to give the Yeoman of England a distinctive and powerful appearance. The tractor and its engine suffered from some reliability problems,

were known as styled versions, distinguishing them from the earlier unstyled models, and traces of the Dreyfuss influence could still be identified in the John Deere range until the new-generation tractors arrived in 1961.

Most of the other leading tractor companies introduced new styling during the late 1920s and the 1930s, with curves replacing the previous angular lines, and there was also a trend towards brighter, more eye-catching colours. Examples included the British-built Fordson Model N, which exchanged its dark blue paint finish for a conspicuous shade of orange as part of a major update in 1937. The colour change was prompted by pressure from the US market, where the salesmen

The influence of the styling specialists was clearly evident when Minneapolis-Moline showed off its new-look tractors in 1938. As well as more rounded lines, tractors such as this 1939 Z series model were given a new bright yellow paint finish which was prairie gold.

and the Yeoman of England survived only a brief production run.

Technical developments such as diesel power, rubber tyres and the three-point linkage with hydraulic control affected the performance of the tractor. They were the most significant improvements during the 30-year period from 1920, but this was also the period when styling became important.

The fact that appearances can influence customers' buying decisions had already brought stylists to the fore in the motor industry, and tractor manufacturers followed the same trend just a few years later. Most of the major styling developments took place in the United States, and some of the biggest tractor manufacturers were prepared to pay big money to hire top industrial designers to give their products a new look. One of the most famous from the 1930s was Raymond Loewy, who was awarded a contract to give the

International Harvester company a new logo and its products a more trendy appearance. His design changes went beyond updating the styling, and included modifications to make the tractors more user-friendly with improved control layouts.

The stylist chosen to give John Deere tractors a facelift was Henry Dreyfuss, another of America's top industrial designers during the 1930s. His brief was to replace the angular lines inherited from the 1920s with a more contemporary rounded appearance. The result was a distinctive new look that first appeared in 1938 on the new A and B models, and later spread throughout the range. Tractors with the Dreyfuss lines

CASE LA

Manufacturer: J. I. Case
Model: LA
Production started: 1940
Power unit: Four-cylinder engine with overhead valves
Power output: 46.6HP maximum
Transmission: Four-speed gearbox, chain and sprocket final drive

were demanding a more up-to-date image for the ageing Fordson.

Massey-Harris introduced a new rounded look for its Pacemaker and Challenger models in 1938. They were called the 'streamlined' models, and an even more rounded look was adopted for other models, including the MH 101. To emphasize the new streamlined styling, a bright red paint colour was also introduced in 1938.

Marketing experts chose special names to make the colours seem more distinctive. One example is the special shade of red adopted in 1939 for Case tractors instead of the previous grey. It was known officially as 'flambeau red'. Allis-Chalmers tractors were painted a special shade of orange after Harry Merritt, head of the Allis-Chalmers tractor operation, noticed large numbers of Californian poppies in full bloom while he was on a train journey. That, he insisted, was the exact colour for the Model U tractor, and it was called 'Persian orange'. One of the brightest of the colour finishes was a sunny shade of yellow chosen by Minneapolis-Moline in 1938 and called 'prairie gold'.

While US manufacturers led the way in the 1930s styling and colour changes, European manufacturers could not ignore the trend. The colour chosen in 1939 for David Brown's new tractor to replace the grey Harry Ferguson preferred was a bright shade of red identified as 'hunting pink', the very English name for the colour of the coats worn when fox-hunting. In 1965, this paint finish made way for a more distinctive white and brown colour scheme, or, as the company insisted, 'orchid white' and 'chocolate brown'.

The LA tractor arrived in 1940 as an updated version of the old model L introduced 11 years previously. LA design features included new curved lines and the 'flambeau red' paint colour first seen in 1939.

POWER ALTERNATIVES

The first tractor fuels were gasoline or petrol, but companies such as Hart-Parr and Rumely gained extra sales by developing engines burning cheaper kerosene or paraffin. The semi-diesel engine was favoured by European farmers because it could burn almost any fuel including waste engine oil. The search for the perfect tractor engine has also covered updated steam engines, gas turbines and electric power. Bio-diesel produced from farm crops and fuel cells may be future options.

Most of the pioneers who built the first generation of tractors in the United States and Europe relied on spark ignition engines fuelled by petrol or gasoline, but there were also those who preferred to use paraffin or kerosene because it was less expensive. Early converts to paraffin included Charles Hart and Charles Parr. The engines they designed used oil instead of water for cooling, and this allowed the higher operating temperatures needed to burn low-grade fuels more efficiently.

The Rumely tractor range was also designed to burn low-grade fuels, and this was strongly emphasized as an important sales feature. The OilPull brand name was a reminder that the engine burned kerosene oil fuel, and the first Rumely tractor built and tested in 1909 was known by the nickname 'Kerosene Annie'. A surviving photograph of one of the first production tractors in 1911 has the slogan

Two big heavyweights from the early twentieth century. Rumely (left) advertised their tractors' ability to burn cheaper, low grade fuels through their OilPull brand name, while the Nichols and Shepard 25-50 model (right) was one of their Oil-Gas tractors.

'Burns Kerosene at All Loads' prominently painted on the side of the cooling tower.

The Rumely engine's ability to run on paraffin was partly due to the higher operating temperature offered by the oil-filled cooling system and partly because of the special carburettor designed for low-grade fuels. The engine design was by John Secor, who had spent many years developing engines capable of burning fuels such as paraffin. The reputation he built up had attracted the interest of the Rumely company, which employed him in 1908 as head of its design team.

Engines running on lower-grade fuels such as paraffin are less efficient and have a lower power output than when burning petrol, but farmers were attracted by the cost savings. Spark ignition engines designed to start on petrol and then run on paraffin after a warming-up period were soon established as the standard power unit for farm tractors, both in the United States and the United Kingdom. This continued until improved high-speed diesel engines began to take over in the 1950s.

AN ALCOHOL ALTERNATIVE

Another option for fuelling spark ignition engines is alcohol, and this has attracted interest in Britain and some other European countries because it can be produced from familiar farm crops such as potatoes or grain. One of the first attempts to use alcohol as a tractor fuel came in 1905 when Dan Albone, one of Britain's leading pioneers of tractor development, measured the performance of one of his Ivel tractors fuelled first with petrol, followed by paraffin and then alcohol.

The test was organized because Albone was concerned about the price of petrol. At that time, Britain was completely dependent on imports to fuel the tiny number of cars and other vehicles, and the price was increasing.

Using a home-produced fuel could be less expensive, he thought, and also more strategically secure. The trial, which was widely publicized in the press, was partly to obtain performance data for alternative fuels. Albone was also shrewd enough to realize that farmers would be likely to take a more favourable view of tractor power if there were a possibility that the raw material for the fuel could be grown in their own fields. To make the comparison, he measured the distance his tractor was able to plough on exactly two gallons (9.09 litres) of each of the fuels. As expected, petrol was the most efficient of the three fuels. Paraffin was in second

The production total for International Harvester's Mogul tractors had reached more than 20,000 when production ended in 1919. They were powered by kerosene-fuelled engines, and heavyweight models, such as this 12-25 with a twin-cylinder engine, used a tower-type cooling system.

place, ploughing only 4.2 per cent less than with petrol in the tank, while the distance ploughed on alcohol was 12.5 per cent less than with petrol.

The idea of filling the fuel tanks of tractors with alcohol has resurfaced from time to time, and the last occasion was during World War II when the British Government recognized the strategic value of having a home-grown fuel supply for some of the essential vehicles such as tractors. It was a time when imported fuel and food supplies were threatened by German U-boat attacks, and there was an official government request for information about how many acres of potatoes would be needed to produce enough alcohol to fuel a tractor for a year's work. There is no evidence that this information was ever put to practical use. It may have

been decided that growing crops for food was more important than using the land to produce fuel crops.

There was a different approach to engine development in some European countries where the semi-diesel or hot bulb engine dominated the tractor market for many years.

THE HOT BULB ENGINE

British engineer Herbert Akroyd Stuart patented the earliest hot bulb engine, and this was the type of engine used by Richard Hornsby, the first British company to build tractors commercially. It advertised a range of Hornsby-Akroyd tractors powered by hot bulb engines in four sizes from 16HP to 32HP, although all of the four tractors they actually built and sold had the 20HP engine.

Rumely's early success was based partly on strength and build quality, and partly on the fact the tractors could burn kerosene or paraffin. This latter ability was promoted through the use of the OilPull brand name, displayed on the side of the rectangular cooling tower of this 30-60 model.

Hot bulb engines were usually a single-cylinder, horizontal design. They were slow revving, and the ignition system was a point inside the cylinder head that was hot enough to ignite the fuel as it entered the combustion chamber. While the engine was running, the combustion process heated the hot point or hot bulb; however, for a cold start the operator used a blowlamp to heat the bulb. The time needed for a cold start was an obvious disadvantage of this type of engine. It also earned a reputation for uneven running, but there were some virtues that made the hot bulb popular in a number of European countries, including Germany, Italy, France and Sweden. One attraction was long-term reliability with little maintenance, and this was mainly due to an uncomplicated design with few moving parts. A simple design also helped keep production costs low, and farmers liked the hot bulb engine's ability to burn almost any type of liquid fuel including paraffin, waste engine oil or even creosote.

Although the hot bulb engine was never popular for powering tractors on American farms or in its country of origin, farmers in Germany were more enthusiastic. The German-based Lanz company was easily the biggest manufacturer of tractors with hot bulb power, and its Bulldog models were popular for more than 30 years from the early 1920s. Other prominent tractor companies in this sector included Munktells in Sweden and Italy's Landini company.

The low efficiency of the hot bulb engine is indicated by the biggest of the Landini tractors from the 1930s. It was called the SuperLandini, and the power output at 620rpm was officially described as '40HP normal or 48HP maximum', produced from a single cylinder with a massive 12.2-litre (744–

This cutaway illustration of a 1930s Lanz Bulldog clearly shows the semi-diesel engine with its large-diameter single cylinder in a horizontal position and the combustion chamber at the front.

cu. in) cubic capacity. The output from the six-cylinder diesel engine of a modern John Deere 7610 tractor peaks at 143HP from 6.8 litres (415 cu. in) or 1.13 litres (69 cu. in) per cylinder.

THE STEAM ENGINE

Steam was comprehensively beaten in the competition to be the prime power source for mechanized farming, but there were some enthusiasts who believed an updated version of the traditional design could give steam a new lease of life for farm work. It seemed an attractive idea, and for a while it was supported by evidence from the car industry in the United States where companies such as Stanley and White had achieved some success with vehicles powered by what was then a high-tech version of the steam engine. A steam car briefly held the world land speed record in the early 1900s. The US steam car manufacturers had shown how most of the disadvantages of the old-fashioned steamer could be eliminated, producing a power unit capable of matching – and in some respects beating – the performance of the internal combustion engine, which was then at a relatively early stage of development and still had plenty of faults.

One of the disadvantages of the traditional steam engine is the need for a supply of dirty, bulky coal, which has to be shovelled by hand to heat the water. Modifying the firebox to burn wood or even straw was useful for areas where coal was difficult to obtain, but these fuels were less efficient than coal and made stoking the fire an even more demanding task. The new steam cars used a paraffin burner to heat the water, reducing the space required for fuel and eliminating the need for someone to feed the fire and remove the ashes. An additional benefit from the switch from solid to liquid fuel was the ability to control the heat quickly and precisely by turning a tap on the fuel line to the burner, providing the driver with a more responsive method of controlling the power output.

Disadvantages of traditional steam power also include the volume and weight of water in the boiler, plus the need for a supply of water to replace the large amounts lost as steam. In the early 1900s, steam engine technology overcame these problems by using a small tubular boiler, allowing a big reduction in size and weight; having a much smaller volume of water to heat also reduced to just a few minutes the time needed to reach full working pressure after a cold start. Fitting an efficient condenser reduced the water loss and allowed a full day's work without a refill, eliminating the need for regular visits from a water cart.

The new generation of steam cars from Stanley, White and others was also able to claim some advantages over petrol-powered rivals. These included the unbeatable smoothness and excellent torque characteristics of steam, plus much lower noise levels. Steam also offered better reliability. The principal problem areas for the early petrol engines included primitive ignition systems and carburettors, and these were absent from steam engines which at that time also benefited from well over 100 years of development.

> ## LANZ BULLDOG
>
> **Manufacturer:** Heinrich Lanz
> **Model:** 15/30 Bulldog
> **Production started:** 1929
> **Power unit:** Single-cylinder semi-diesel
> **Power output:** 30HP
> **Transmission:** Three forward gears with 6km/h (3.7mph) top speed

The words 'Advance Rumely' on this tractor show that it was built after 1915. A sudden downturn in 1914 forced the Rumely company into bankruptcy, and it was re-established in 1915 as the Advance Rumely Thresher Co.

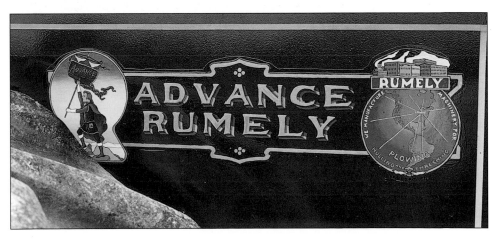

This is the first production version of the Heinrich Lanz company's first Bulldog model. Production started in 1921, using a single-cylinder semi-diesel engine developing 12HP. The wheels were equipped with solid rubber tyres, and the tractor was designed mainly for stationary work.

Although the Bryan tractor was designed as a general-purpose power unit for drawbar work and powering a belt pulley, the emphasis in the publicity was on the benefits of a steam engine for driving the belt to power a thresher. There was also a small amount of added nostalgia. 'It's good old steam power modernized' read the sales leaflet reassuringly.

While the steam car was enjoying a brief period of success in the United States, manufacturers of tractors and agricultural steam engines ignored the new technology, waiting until the early 1920s before making a serious attempt to adapt it for farm work.

THE BRYAN STEAM TRACTOR

The most successful steam tractor development project came from the Bryan Harvester Co. of Peru, Indiana, with tractors available commercially for about four years from 1922. It was, according to the Bryan sales leaflet, 'The World's First Light Steam Tractor'. The power output was rated at 20HP, and, with a length of 3.56m (11ft 8in) and width of 1.96m (6ft 5in), it was about the same size as a conventional tractor with similar power. The 2500kg (5500lb) weight was also comparable with a petrol/paraffin tractor, and the 273-litre (60-gallon) capacity of the water tank was said to be sufficient for a full working day, with an efficient tubular condenser to recycle water after the steam had passed through the cylinders.

The burner was designed to work with paraffin or any grade of distillate,

and the fuel tank capacity was 136.5 litres (30 gallons). The fuel system included an automatic control to cut the supply to the burner when the boiler pressure reached the 272kg (600lb) maximum, and there was a pilot light to reignite the main burner when the fuel supply was restored. The boiler consisted of a series of tubes which were 'Absolutely Non-Explosive', the leaflet stressed – although it failed to explain why the tubes were immune from exploding. Steam from the boiler powered a pair of pistons with 101.6mm (4in) bore and 127mm (5in) stroke, operating at a leisurely 220rpm rated speed. The range of travel speeds from the single-gear transmission altered with the engine speed, allowing infinitely variable adjustment up to the km/h (5mph) maximum.

Twin City was the brand name of the Minneapolis Steel and Machinery Co., which competed successfully in the heavyweight sector of the North American market. The first 40 series tractors were built in 1911; the design was updated in 1913 when the canopy was extended forwards.

'There can be no question about steam being the ideal power for belt work. For years practically all of the grain threshing of the world has been done with this power. Despite the fact that it is necessary to haul coal and water to the steam traction engine, practical men have preferred it to any other kind of power. The Bryan Light Steam Tractor has all the advantages of the steam traction engine in stored and steady power - dependable and flexible power with none of its disadvantages as regards fuel, water and weight.'

In spite of the benefits claimed for the Bryan tractor, sales were disappointing, and this fact may have influenced International Harvester when it decided to abandon its steam tractor development programme. International Harvester started experimenting with steam tractors in about 1920 and, by 1924, it had built several different prototype models which, like the Bryan version, were based on a paraffin-fuelled burner heating water in a tubular boiler. The International engine was a V-twin design, with the cylinders located beside the driver's seat and covered with a thick layer of insulation to reduce heat wastage, and it was linked to a condenser to recycle the water.

Above: This Twin City 40 badge is positioned on the cylindrical radiator at the front of the TC-40 tractor. Production of the 40 model with its 40-65HP rating continued into the early 1920s, and it was among the first batch of tractors tested at Nebraska in 1920.

This is the first tractor built by Munktells of Eskilstuna, Sweden's biggest manufacturer of agricultural steam engines in the 1890s. It diversified into tractor production in 1913, using a twin-cylinder, semi-diesel engine with 40HP maximum output. Production continued for two years.

Britain was the traditional home of steam power. The steam engine was invented in Britain, and British engineers were responsible for most of the development work to adapt steam power for industry, transport and agriculture. When tractor power began to challenge the success of the traction, portable and ploughing engines on farms, some of the steam engine manufacturers – including Ransomes and Marshall – were quick to experiment with tractor designs of their own. Others, however, simply continued to build their traditional steam equipment for a rapidly shrinking market.

ATTEMPTS TO MODERNIZE STEAM
There were also some attempts to modernize steam power and bring it into the tractor era. This was the approach adopted by Garrett, one of Britain's oldest farm equipment manufacturers. The company was established at Leiston, Suffolk, in 1782, and it was already more than 130 years old when work started in about 1915 on a new project to develop what was later called the Suffolk Punch, named after one of the most popular working horse breeds.

This was a steam-powered tractor, and the design team started with a completely clean sheet of paper, moving the driver from his traditional position at the rear of a traction engine to a seat above the front wheels. The position of the boiler was also reversed, with the firebox facing forwards, and this moved most of the weight of the boiler to a position above the driving wheels at the rear. An auxiliary water tank was also located over the rear axle to add extra weight. Other design features included a double-crank compound engine producing 40HP, and moving the driver forwards allowed the designers to use Ackermann steering. Similar to that used on some trucks, this type of steering was much lighter and more precise than the chain-operated steering traditionally used on traction engines.

The emphasis on weight over the rear

wheels was aimed at improving grip to provide maximum traction, as the Suffolk Punch was designed for ploughing and other pulling work as well as for threshing, and the test programme for the first production tractor in 1917 was based on ploughing. Although the Garrett tractor lacked some of the latest high-tech steam features of the Bryan and International tractors, it was an ambitious attempt to modernize traditional traction engine design. In commercial terms, however, it was a failure. Just eight of the steam tractors were built, and one of these – called the Joker - has survived.

The Suffolk Punch, like its US equivalents, probably arrived too late. Steam was increasingly regarded as outdated, while new arrivals such the Fordson Model F focused attention on cheap, versatile tractor power, and progressive farmers with money to invest in a tractor wanted the new technology rather than the old. With the advantage of hindsight, the decision by Garrett to use the Suffolk Punch name may also have been a mistake, as the

Nichols & Shepard is another of the great names from the early days of American farm equipment history. Tractor production started in 1911, but steam engines were still being built by the company in the mid-1920s, long after most of its rivals had pulled out of the market.

association with the slow pace of a carthorse may not have helped its new tractor's image.

EXPERIMENTING WITH ELECTRICITY

Experiments with electricity for powering field machinery are almost as old as the tractor, and there have been numerous attempts to use electric motors in tractors and cable ploughing

equipment. One of the first companies to demonstrate a cable ploughing system using electricity was F. Zimmermann of Halle, Germany. A report published in 1892 in an English trade magazine suggested that a public demonstration of the equipment had been successful, and some cable sets had already been sold. The 1892 version used an electric motor to power a winding drum anchored at the headland of the field, with a steel cable spanning the field to another anchor on the opposite headland. A two-way plough attached to the cable was pulled to and fro between the drum and the anchor points, which had to be moved to a new position each time the plough completed a crossing. It was obviously based on the steam-powered cable ploughing systems developed in Britain by John Fowler and others, but moving the electric

some of the methane from the digestion process has to be burned to maintain a suitable temperature inside the digester, and this reduces the amount of methane available for other uses.

One of the attractions of the anaerobic digestion process for methane production is that it can operate with reasonable efficiency on a small scale. A digester using the manure from a pig-fattening unit or intensively housed poultry could, in theory at least, fuel a tractor with enough left over to heat the farmhouse. The tractor fuel is carried in pressurized cylinders filled from the storage unit; portable digesters small enough to fit on a tractor could generate gas while the tractor worked.

A British engineer who studied several methane installations in Germany in 1951 wrote an enthusiastic report in the magazine *Farm Mechanization*. One of the farms he visited had 90 hectares (220 acres) of crops and grass. Here, the manure from 55 cattle of mixed ages, 180 pigs and 650 poultry produced 42,500 cubic metres (1.5 million cubic feet) of gas per year, said to be equivalent to 30,000 litres (6600 gallons) of petrol. One-third of the gas was used to heat the digester and power various pumps associated with the plant and the storage container, and the rest powered two tractors with 28HP engines and also provided the house with fuel for

heating and cooking. The sludge remaining after the digestion process has finished is used as a manure with virtually undiminished fertilizer replacement value, according to the methane enthusiasts.

ALTERNATIVES TO METHANE

Although there are occasional bursts of enthusiasm for farm-scale methane production, commercial acceptance is still slow and there are a number of significant problems. One of these is the additional management expertise required to run the plant successfully. Also, the process is less suitable for dairy and beef farms where manure is not available for much of the year while the cattle are grazing. Another difficulty is that, when methane is produced for domestic heating, the period of maximum demand in winter is also the time when more gas is needed to maintain the required digester temperature.

Producer gas is an alternative to methane. It consists of a mix of carbon monoxide, methane, hydrogen and oxygen, and it is generated by passing steam over red-hot carbon. The carbon is usually made from wood, which is readily available in most areas, helping to encourage the popularity of the kits in wartime Europe and currently in some developing countries.

Modern producer gas kits for mount-

ing on tractors include a combustion chamber with a fire grate and burner, plus various scrubbers and filters to prepare the gas as a fuel. France, probably the country that made the biggest use of producer gas to fuel vehicles during World War II, still produces a modern version of the tractor kits for export. According to the manufacturers, the starting-up time from cold is about 10 minutes, with a further 10 minutes needed each day for other routine chores, including ash removal. The gas can be used as 100 per cent of the fuel for a spark ignition tractor engine, but the limit for a diesel engine is about 70 per cent producer gas and 30 per cent diesel. In both cases, the power loss is likely to be at least 20 per cent.

GAS TURBINE TECHNOLOGY

Probably the most eye-catching of the tractors designed to demonstrate a different type of power unit or an alternative fuel is the HT-340 experimental model announced in 1961 by International Harvester. It was described at the time as a research vehicle designed to investigate the possibilities of using a gas turbine as a power unit for farm equipment. The sleek, futuristic styling of the tractor, plus the fact that the International Harvester driver at the press launch was apparently wearing a motor racing type helmet, suggests that publicity was also high on the list of priorities for the HT-340.

Gas turbines were big news during the 1950s and early 1960s. The Rover company in Britain had gained massive publicity when they demonstrated the world's first gas turbine-powered car, and they gained more headlines when an updated version of the car competed successfully in a special section of the Le Mans 24-hour race in France. Most of the other leading car companies in Europe and the United States followed the lead from Rover and were involved in some form of gas turbine research. There were plans from the Chrysler company to build a pre-production

batch of about 50 cars with turbine power for a large-scale evaluation programme, including some to be loaned to members of the public. Gas turbines were also being used in power stations and to power aircraft and ships.

International Harvester became involved in gas turbine technology because one of its subsidiary companies was developing a turbine to power a small helicopter, and a de-rated version of this power unit was used for its tractor research programme. A gas turbine burns fuel to heat a stream of air flowing through a combustion chamber. The air expands as it is heated, and the expansion increases pressure in the chamber and forces the air to accelerate towards the exit at the rear of the engine. On its way, the fast-moving air stream passes through the blades of a turbine, forcing the turbine to rotate.

This form of power unit has a number of advantages. Turbines are extremely smooth in operation, and they are also

highly responsive – which can mean good acceleration. They are relatively light and compact in relation to their power output, they have a good reputation for long-term reliability and they can burn a wide range of fuels, including paraffin.

A small, lightweight power unit with super-smooth operation, good acceleration and the ability to run on lower-cost

ALLIS-CHALMERS

Manufacturer: Allis-Chalmers
Model: Experimental fuel cell tractor
Production started: 1959
Power unit: Electric motor powered by 1008 fuel cells
Power output: 20HP
Transmission: Gear drive forwards; electric motor reverses to travel backwards

Allis-Chalmers recognized the potential of fuel cell power in the 1950s, and it built several experimental tractors using fuel cells made by another company within the Allis-Chalmers group. This 1959 tractor uses a bank of 1008 cells to power a 20HP electric motor.

SPECIAL TRACTORS FOR SPECIAL JOBS

Orchard work demands a tractor with a smooth outline to avoid damaging branches, and this is just one of the special types of tractor manufacturers have produced. There have also been transport tractors designed for load carrying, plus tractors designed with a built-on powered cultivator. Another special category is high-horsepower heavyweights developed specifically for the big-acreage farms, and the leader so far in this sector is Versatile's Big Roy eight-wheeler.

When Dan Albone was developing his Ivel tractor in the early 1900s, his aim was to make it as versatile as possible, but even he was forced to accept that there are some farming situations which demand a special type of tractor. The first special version of the Ivel was built in 1903 for export to Australia, where a fruit grower in Tasmania had asked for what was described as a 'low-profile' tractor to be used for grass-cutting work in his orchards. Exactly how Albone modified the Ivel for working under the tree branches is not clear, as no photographs have survived, but it was almost certainly the world's first orchard tractor.

When the Versatile company decided to build the world's most powerful tractor, the result was Big Roy, a 600HP giant. However, there was no farm machinery big enough to make use of this power, so Big Roy never passed the prototype stage.

While Dan Albone's orchard tractor was probably a one-off product, many of the leading tractor companies in the 1920s and 1930s were making specially modified tractors to meet the needs of fruit growers.

The demand for this type of tractor was particularly strong in the United States, where the production of tree fruit and nuts was expanding rapidly. Typical features of the tractor conversions for working in orchards included removing vertical air cleaner stacks and exhaust pipes, and tucking them away below the bonnet line to reduce the overall height of the tractor, thereby avoiding tangling with overhead tree branches.

THE ORCHARD TRACTOR MARKET

Massey-Harris produced an orchard version of its mid-range Pacemaker tractor, available on the unstyled version from about 1936, and this model illustrates some of the other modifications offered to fruit growers. The usual spoked design of tractor wheels in the 1920s and 1930s could damage branches, so either a protective disc covered the front wheel spokes or, as on the Pacemaker, solid wheels were fitted to the front of the tractor.

The rear wheels were even more of a hazard, partly because of the spokes, but also because of the spade lugs or cleats sticking out from the rims. The usual answer to this problem, again illustrated on the Pacemaker, was to extend the rear mudguards or fenders to cover the upper half of the wheel completely, while the front section of each fender was extended forwards to push branches aside with little or no damage.

Mechanical specifications were usually the same on the standard and orchard models, but one alteration offered on some orchard models, including the Pacemaker, was the addition of independent steering brakes on the rear wheels in order to allow a smaller turning radius and improve the manoeuvrability.

While most of the big US companies offered orchard models, Deere & Co. was particularly active in catering for this sector of the market. It produced a

---

PACEMAKER

Manufacturer: Massey-Harris
Model: Pacemaker
Production started: 1936
Power unit: Massey-Harris four-cylinder, water-cooled
Power output: 27.5HP maximum
Transmission: Four-speed gearbox

The demand for standard tractors is much bigger than for orchard models, and this certainly applies to the orchard version of the Massey-Harris Pacemaker. Its curved U-frame enclosing the underside of the engine and gearbox shows clearly in this side view of the Pacemaker orchard model.

series of specially modified tractors that are now prized by John Deere enthusiasts because of their distinctive design features.

JOHN DEERE

John Deere orchard models were identified by the letter 'O', and the series started with the orchard version of the General Purpose or GP model in 1931, identified as the GPO model. As well as the usual repositioned air inlet and exhaust pipes and the redesigned rear fenders, some of the later GPO tractors were also supplied with a redesigned seat support that resulted in a lower driving position to reduce the risk of contact with overhead branches. Production of the GPO ended in 1935 with fewer than 600 built.

Some of the GPO tractors were shipped to Yakima, Washington, where the Lindeman brothers fitted them with

a set of tracks. The Lindemans were the local John Deere dealers in an important fruit-growing area. Some of the apple orchards were on steep land, and the owners wanted orchard tractors equipped with tracks to provide extra stability. Only a small number of the GPO tractors were equipped with tracks, but the number of conversions increased sharply when the orchard version of the John Deere Model B tractor arrived. Deere supplied about 2000 of the BO tractors to Yakima for the track conversion; in 1947, the Lindeman business was purchased by Deere to become a specialized design and production centre for John Deere crawler tractors.

The orchard versions of both the Model B and the Model A tractors were introduced in 1936. The standard B tractor had recorded a maximum output of 15HP in the belt tests at Nebraska in 1934 and the figure for the Model A was almost 25HP. The Model A resulted in one of the most spectacular-looking production models of John Deere's two-cylinder series. This was the AO Streamlined (AOS) model, built in small numbers from 1937 to

The rear view of the orchard Pacemaker shows how the profile of the tractor is kept low, with tall structures such as an upright exhaust pipe eliminated. Independent brakes added to each rear wheel to allow the tractor to be turned at a sharper angle at the end of a row of trees or bushes.

1940 and available with either steel wheels or rubber tyres. The rear wheels designed for the rubber tyres were solid, with no spokes to tangle with tree branches. It earnt its 'Streamlined' tag because of its smooth shape, designed to allow the tractor to slide past tree branches with little risk of damage to either the tree or the driver. The AOS tractor was designed for orchard work, from the specially shaped grille protecting the radiator to the rear fenders which were extended forwards past the engine compartment. At the rear of the tractor, the seat was designed to keep the driver as low as possible. The overall height of the AOS was only 1.35m (53in), and the width was reduced to just under 1.42m (56in) to make the tractor narrow enough for vineyard work.

Another version of the AO tractor arrived in 1949 based on the styled

drive was by a roller chain and sprockets to a differential on the live rear axle. The rear axle was solid, with one wheel keyed onto the axle and the other mounted on a sleeve over the axle.

Britain's Daimler company built its reputation by making luxury cars for a long list of prestigious customers, including the Royal Family, and it caused some surprise when it decided to add farm tractors to its product range. The company's first tractor arrived in 1911 and it was, of course, a very up-market model. The 30HP, four-cylinder power unit was produced for one of its cars and featured the patented sleeve valve design that made Daimler engines among the quietest in the motor industry.

The gearbox on the Daimler tractor provided three forward speeds and a reverse, and the drive was transmitted through a leather-faced cone clutch. A rear-mounted pulley was provided for belt work, and there was an automatic locking device to hold the gears securely in neutral while the drive to the belt pulley was engaged. An unusual feature inherited from the car side of the

Daimler business was a plunger pump linked to an automatic lubrication system. As most Daimler drivers were accustomed to life's luxuries, the tractor was equipped with a canopy to provide weather protection for the driver. Also included in the standard specification was a load box, but, as the company made only a passing reference to it in its publicity material, the transport function appears to have been of secondary importance to the drawbar and belt-pulley applications.

ANDRE CASTELIN

Although the majority of French engineers appear to have been intent on reinforcing their country's early successes in car design and production, André Castelin was one of the exceptions. He was more interested in tractor development, and he had some interesting but unconventional ideas. He was particularly concerned about the damage caused by tractors when working on soft ground, and his aim was to keep the tractor as light as possible in order to reduce soil compaction and

This orchard version of a styled John Deere A tractor, identified as the AO model, was built in 1953. Its metal shielding is even more extensive and provides even more protection for the rear wheels than on the GPO model illustrated on the previous page, which was built about 20 years earlier.

subsequent drainage problems. Even a light tractor could pull equipment such as mowers and binders, but, for heavy-duty pulling jobs such as ploughing, Castelin fitted a powered winch and a cable to enable a light tractor to do the work of a much heavier machine. Using a cable system to pull ploughs allowed the tractor to stay on the headland, and this was more efficient, M. Castelin claimed, than using fuel to take the tractor and driver to and fro across the field with the implement.

Castelin's first prototype tractor was completed in 1900, and he demonstrated it on a farm near Meaux, in France, where it pulled a binder. He built a second, improved tractor in 1904, using the 12HP petrol engine designed for a de Dion Bouton car. This was the load-carrying version with a platform over the rear wheels, and it was designed to carry 1 tonne (1 ton). The winch for the cable was located under the load platform, and the addition of a large and prominent water tank suggests that Castelin had to provide extra cooling capacity to deal with an overheating problem when the car engine on the tractor was used for stationary or slow-speed work. There is no evidence that Castelin achieved any commercial success with his tractor ideas.

LOAD CARRYING IN THE UNITED STATES

While British and French load-carrying tractors were selling in small numbers – or in some cases not selling at all – the Avery company in the United Sates produced what was probably the first tractor of this type to achieve significant commercial success. It was designed as a transport tractor that could be used as a load-carrier on the farm and for road haulage as well, and it was available

from 1909 until production ended in about 1914. The Avery company based at Peoria, Illinois, started out as a successful machinery manufacturer, but it achieved even more success when it moved into the steam traction engine business during the 1890s, establishing a reputation for high-quality engineering, including its famous undermounted engines. The 'farm truck', as it was called, was Avery's first venture into the tractor market and it chose an unconventional design.

Load-carrying tractors were based on a truck layout, and this was a feature of the Avery design, with the load platform at the rear and the driver and power unit at the front. Special features to match its transport role included leaf-spring front suspension, and the transmission included a high ratio top gear giving a 24km/h (15mph) maximum speed on the road. The farm truck's credibility as a tractor was helped by the addition of a belt pulley at the front for stationary work, and there was a

drawbar at the rear for towing implements.

When the farm truck was used on the road, the wheels could be fitted with solid rubber tyres; for field work, there were special steel rims with cup-shaped indentations in which specially shaped hardwood pegs could be inserted to improve traction on a soft surface. The wooden pegs were probably not a success. In wet conditions, the wood was likely to expand and grip the sides of the cups more securely; in dry weather, the reverse was likely to happen, and it must have been very easy to lose some of the pegs. The wooden peg idea was not copied by other manufacturers, and, when Avery later introduced a range of

The 'streamlined' version of the AO tractor, listed as the AOS model, was built between 1937 and 1940. The distinctive styling includes a reshaped radiator grill designed to brush tree branches aside, and its rarity and unusual appearance make it a special favourite with John Deere enthusiasts.

more conventional tractors, they were equipped with steel lugs or cleats.

Avery entered its farm truck in the tractor trials held near Winnipeg, Canada, in 1909. According to the results table, the tractor weighed 2270kg (5000lb), making it the lightest entry in the competition. Avery also had the heaviest entry in a different class at the same event, and this was one of its steam traction engines, which registered a hefty 18,600 (41,000lb) when it was driven over the weighbridge. The Avery farm truck, powered by a 36HP engine with four cylinders, came second against two other entries in its class, and the Avery traction engine came third in a class of five heavyweights.

THE UNIMOG

Although the Avery load-carrier achieved some success, the commercial breakthrough for this type of tractor did not arrive until the tractor boom that followed World War II. The tractor that made the breakthrough was the four-wheel drive Mercedes-Benz Unimog. Albert Friedrich designed it as a new type of load-carrying agricultural tractor, and it was available in small numbers until the project was taken over by Daimler-Benz. The company transferred it to its factory at Gaggenau, where it was able to increase the production volume to keep pace with a big increase in demand.

The original Unimog was a cross between a farm tractor and a transport

BADGES AND TRADE NAMES

Trade marks can be extremely valuable and some of the most famous badges from the motor industry have also appeared on tractors. The Ford oval, the diamond-shaped Renault emblem and the Mercedes-Benz three-pointed star are all familiar on farm tractors.

Other well-known badges in the tractor and farm machinery industry include the leaping-deer emblem for John Deere, the four-eyed tiger badge on the front of Same tractors from Italy and the black-bull emblem used by Same's stablemate, Lamborghini.

If tractor enthusiasts could vote for their favourite trade mark, the eagle emblem once used by Case would be high on the list. It is based on an American bald eagle used as a mascot for one of the companies in the 8th Wisconsin Regiment in the Civil War. Case, also Wisconsin based, adopted Old Abe as its trade mark from 1865 until it was replaced by a new, less distinctive logo 104 years later.

vehicle. It was powered by a 25HP Mercedes engine, and the emphasis was on farm work. Features aimed at the farming market included a power take-off and hydraulic linkage at the front as well as at the rear – the first time this had been available on a farm tractor. The gearbox offered a range of slow speeds for field work as well as faster speeds for travelling on the road, and the four-wheel drive system through equal-sized wheels provided efficient traction to cope with difficult ground conditions. Early versions were also fitted with a pulley for stationary work with a belt drive.

There were some limitations, and one example of these was the relatively small wheel diameter, which was not ideal for ploughing. Despite this, farmers and

contractors were a big percentage of the sales total in the early years of Unimog production. Apart from its obvious transport role, the load platform was also used to boost efficiency for some field jobs. It could carry a load of fertilizer to make it easier to refill the rear-mounted spreader, and bags of seed on the platform were handy for topping up the drill hoppers carried on the rear linkage. Early versions of the Unimog were used for a wide range of field jobs and were also popular for forestry work, and, in Germany particularly, there were many farmers who became Unimog enthusiasts.

Since then, a continuing series of design improvements have tended to put more emphasis on the Unimog's non-farming role. Instead of the 25HP of the original version, engines of 200HP plus are now available in some of the current models. The maximum travel speed has gone up from 50km/h (31mph) in the late 1940s to more than 80km/h (50mph), payloads of 9 tonnes (9 tons) plus are available on some models and levels of comfort and refinement have steadily improved.

A feature added to later models designed by John Deere for fruit growers was a curved metal shield extending back over the instrument panel, seen here on an orchard version of a 60 tractor. It prevented tree branches from contacting the steering wheel and protected the driver's hands.

These developments have not reduced the Unimog's work capabilities as a farm tractor, but they have made it more expensive, and they have also increased its appeal as a military and transport vehicle. Agricultural users, including contractors who use the load space to carry equipment needed to turn the Unimog into a highly efficient crop-spraying vehicle, account for less than 10 per cent of the sales total. Most of the current demand comes from military sources, local authorities and a long list of industries including construction and quarrying.

THE OPPERMAN MOTORCART

Britain's entry in the mid-1940s transport tractor market was the Opperman Motocart. It was developed in 1945, the same year as the Unimog, but was less versatile, and its production life was very much shorter. The Motocart was based on a prototype version designed and built by a farmer who wanted to replace some of the horses he was using for transport work. The idea was adopted by the Opperman company at Borehamwood, Hertfordshire. It was powered by a small 8HP air-cooled engine, usually a JAP, which was attached to the right-hand side of the single front wheel. The power from the engine was taken through a chain and sprocket drive and a clutch to the four-speed gearbox, and this formed a compact unit with the front wheel.

In the late 1940s, there were still plenty of farmers using horses for transport work, and the Opperman company's publicity for the Motocart was aimed at the horse owners. The load space on the Motocart was designed to carry up to 1.5 tonnes (1.5 tons) at speeds of up to 19km/h (12mph). The work rate for haulage jobs around the farm was said to be three times as fast as a horse and cart for a 6.83 litres (1.5 gallons) per day average fuel consumption. As well as selling to farmers, the Motocart also attracted some local authority and industrial customers, but demand dwindled in the face of competition from tractors that were more powerful and offered greater work potential. Production of the Motocart ended in about 1952.

MOTOR CULTIVATORS

Another specialist sector of the market was the motor cultivator, which uses the engine to power a rotary spading or cultivating machine permanently attached to the tractor unit. Powered

Another model in the orchard tractor series from John Deere is the orchard version of the 620. It inherited the hinged steering wheel guard of the previous orchard model based on the John Deere 60 tractor. Production of the 20 series tractors ran from 1956 until 1958.

engine-cooling systems on tractors available at the time the competitions were held. Open systems, in which much of the heat was removed by evaporation, were still favoured by many of the manufacturers, especially for big tractors. It must have been easy for drivers to underestimate how much water would be needed for a full day's ploughing in some remote field on a prairie farm.

One of the tests at the 1911 competition was measuring the performance of the tractor on a dynamometer while it delivered its maximum power for 30 minutes. This is a tough test of cooling systems because the engine is working hard continuously and is likely to be producing plenty of heat for the full period. The organizers were obviously aware that the water consumption figures would be of great interest to many potential tractor buyers.

There were 25 tractor entries in the 1911 competition, and six of these recorded no water consumption during the test – presumably all of them had closed cooling systems. The thirstiest of the petrol-powered tractors was the Kinnard Haines, with a 29.4HP average power output during the test and 11.4 litres (2.5 gallons) water consumption during the 30 minutes. The average consumption figures were much higher in the section for paraffin or kerosene tractors, where the four International Harvester entries were all conspicuously

thirsty with consumption figures averaging more than 13.6 litres (3 gallons) for the 30-minute test. The biggest drinker in this section, however, was the new OilPull Model E 30-60 tractor from Rumely.

Power output from the Rumely averaged 68HP and the water consumption was 97.8 litres (21.5 gallons) in the 30-minute test. Exactly why the Rumely tractor required so much water is not clear, but it is possible that a leak in the

cooling system may have caused some of the loss. Rumely also had a smaller tractor entered for the 1911 competition, and this model with a 32HP output lost only 6.4 litres (1.4 gallons) of water in the 30-minute test, one of the best performances in its section.

The big drinkers in the results tables for the Winnipeg competitions were, of course, the steamers. The consumption figures for the same 30-minute test ranged from about 846 litres (186 gallons) for the Canadian-built Sawyer-Massey with 118HP average output to 1574 litres (346 gallons) for the Avery entry, the most powerful engine in the steam section with 159HP average output. The steam engines also consumed large amounts of coal during the half-hour test, ranging from 127kg (279lb) for the Canadian-built Sawyer-Massey to 236kg (543lb) for the Avery. Extravagant consumption of both water and coal is

Big-acreage farms in North America provided a good market for heavyweight tractors with plenty of pulling power during the 10 years from about 1905, and Hart-Parr was one of the companies that built tractors for this sector. One of its most popular models was the 30-60 tractor.

Gaar Scott of Richmond, Indiana, switched from steam traction engine production to tractors in 1910, starting with the petrol-powered 40-70 model, which competed in the 1911 Winnipeg trials. In the ploughing section, the 40-70 achieved the fastest work rate.

one of the obvious reasons why steam traction engine sales were beginning to fall rapidly in the face of competition from big tractors such as the Model E.

RUMELY AND HART-PARR

Rumely's 13.2-tonne (13-ton) Model E tractor remained in production until the early 1920s, and it was tested at Nebraska in 1920. The water consumption was still higher than most of its rivals, but the work capacity of the tractor was demonstrated by the drawbar pull figure in the tests. This reached a maximum of 49.9HP compared with Rumely's very conservative 30HP rating. For farmers with a big acreage to cultivate, the Model E was one of the most popular and durable of the high-horsepower tractors, but it was certainly not short of rivals.

Hart-Parr also built a 30-60 tractor and, like its Rumely rival, it earned a

BIG ROY

Manufacturer: Versatile
Manufacturing Co.
Model: Big Roy
Production started: 1977
Power unit: 19-litre (1159 cu.
in) Cummins diesel
Power output: 600HP
Transmission: Six-speed gearbox
and eight-wheel drive

of high-horsepower tractors. The tractor's official name was the Versatile 1080, but to most people it was known as 'Big Roy'. The tractor was first built in 1977 and was named after Roy Robinson, the (6ft 4in) tall principal shareholder in Versatile. He was a larger-than-life character who, it is said, often arrived at his office wearing a Stetson hat and cowboy boots.

The 1970s were a boom sales period for high-horsepower tractors in North America. 'High horsepower' at that time meant tractors in the 250 to 350HP sector, but there were forecasts that power outputs would continue to increase as large-acreage farmers looked for faster work rates and increased efficiency.

Versatile, as a leading manufacturer of big tractors, had to be ready to respond to demands for more power, and Robinson instructed his engineers to build the biggest tractor in the world.

They chose a 19-litre (1159-cu. in) Cummins engine designed to deliver 600HP, but there were concerns about the problems of converting this amount of engine power into drawbar pull. With an ordinary four-wheel drive layout, the engineers were worried about damage to the final drive and to the tyres, and there was also a risk that the weight of such a heavy tractor would cause soil compaction if it were carried on only four wheels. The design team produced two prototypes because of the uncertainties. One of them split the drive through three powered axles with six-wheel drive, and the other had eight-wheel drive through four powered axles. The eight-wheeler was kinder to the soil, spreading the weight over a bigger surface area, and the four powered axles also offered more security when handling the full 600HP output from the engine.

The eight-wheeler also looked more impressive and attracted more interest, and this is why Big Roy survived while the less fortunate unnamed six-wheeler was scrapped.

This is Big Roy, the result of the Versatile company's ambition to build the world's most powerful tractor to meet the needs of big-acreage farmers. The design team chose a 600HP Cummins engine, which they mounted high up in the rear section of the pivot-steer tractor. Eight wheels turned the engine power into drawbar pull.

Big Roy, now preserved in the Manitoba Agricultural Museum at Brandon in Canada, weighs 26.4 tonnes (26 tons) and measures 10m (33ft) from front to back. Built in two halves, with a hinge point at the centre for the articulated or bend-in-the-middle steering, it has a number of features that were considered to be state of the art in the 1970s. The cab is fully air-conditioned, considered a rare luxury when the tractor was built, and there is a screen inside the cab showing images from a closed-circuit television camera built into the rear of the tractor. The camera is necessary because the driver has virtually no rear visibility from the cab, and the pictures allow the tractor to be reversed more safely and also help the driver to position it correctly for hitching up to trailed implements.

When Big Roy was completed, it attracted a huge amount of press publicity and there were numerous

requests for demonstrations and from farmers willing to help in the field test programme – and this is where the problems started. The obvious problem was finding implements that were big enough to provide any sort of challenge to the world's biggest tractor. A 20-bottom plough working at 305mm (12in) depth was no problem for Big Roy, and the 600HP engine made light work of a 36.6m (120ft) wide cultivator pulled at speeds of up to 10.5km/h (6.5mph), leaving plenty of power in reserve. It soon became obvious that farmers would not be willing to buy the tractor unless the implement manufacturers could be persuaded to make a new range of equipment to suit

this amount of pulling power. Tyre damage was a problem because of stress on the casing and lugs caused by scuffing when the tractor was making sharp-angle turns under power, and the steering system also needed redesigning to provide additional stability as the tractor was turning.

Big Roy was displayed on the Versatile stand at many of the leading North American agricultural shows, and this plus the international press coverage provided at least some return for the large amounts of money Versatile had invested in the project. The tractor is now, quite literally, one of the biggest attractions at the museum at Brandon, just a short drive from the Winnipeg

Manoeuvrability is not Big Roy's best feature, and the overall length means the tractor needs plenty of turning space. Because of the limited rear visibility from the cab, a video camera was mounted at the rear of the tractor and linked to a screen in the cab to help when reversing or hitching up to an implement.

factory where it was built. Having abandoned the Big Roy development, Versatile chose a different project, a four-wheel drive tractor with articulated steering. It was powered by a more modest 470HP version of the 19-litre (1159-cu. in)Cummins engine, and it arrived on the market as the Model 1150 tractor.

POSTWAR EXPANSION

The end of wartime restrictions in the late 1940s heralded a period of massive expansion for the tractor industry, and sales boomed as more farmers switched from animal to mechanized power. Postwar prosperity also encouraged the introduction of technical improvements, including four-, six- and even eight-wheel drive machines, as well as high-speed diesel engines. These were later followed by high horsepower models, systems tractors and bi-directional designs.

The end of the war in 1945 was the start of a period of prosperity and expansion for tractor manufacturers. Tractor power was urgently needed to help much of Europe's farming industry recover from the devastation that had occurred during World War II, and there were still large numbers of working horses to be replaced by tractor power on both sides of the Atlantic. Farm mechanization was also one of top priorities in the Soviet Union and among its Eastern European allies, and large amounts of state funding were provided to establish or expand tractor production in Bulgaria, Czechoslovakia, East Germany, Hungary, Poland and Romania, as well as factories in the Soviet Union.

When the Ford NAA tractor arrived in January 1953, it was known as the Golden Jubilee Model to celebrate 50 years of Ford tractors. As well as a fancy badge, NAA features included a new overhead valve engine developing 30HP from 2200cc (134 cu. in).

The factory at Minsk, in Belarus, was established in 1946 to build crawler tractors, and the first wheeled models began arriving in 1953. It soon became the world's biggest manufacturer, building its millionth tractor in 1950, with the production total passing the two million mark in 1984 and reaching three million in 1995. The success of the Minsk factory and its Belarus tractors was helped by the vast area of agricultural land in its home market and almost complete protection from imported competition; however, the Belarus range also attracted substantial export orders prompted by its extremely low prices. The Minsk factory and other Soviet tractor plants, including those at Kharkov, Kirov, Lipetsk and Volgograd, may have relied mainly on their own technology and built tractors that generally lagged behind their Western rivals in terms of design and sales features, but their products also earned a reputation for rugged performance.

Some of the Eastern European factories attempted to reduce the technology gap by buying designs from manufacturers in the West, and tractors from Massey Ferguson's Banner Lane factory in England were particularly popular. The Polish Ursus factory has used both Zetor and Massey Ferguson designs, and its current tractors still retain Massey Ferguson styling features, while the IMT tractor plant in the former Yugoslavia has based its tractors on Massey Ferguson designs since the 1950s. Deutz-Fahr tractor designs from Germany were also adopted for Yugoslavian production under the Torpedo brand name, and Fiat-based tractors were produced in Romania under an agreement with the big Universal (UTB) tractor plant.

EAST–WEST SHARING IN HUNGARY

West-to-East technology deals in the tractor industry were far from new. One of the first tractors built in Poland was a version of the International Titan 10-20 assembled in the early 1920s by the Ursus company. Immediately after World War II ended in 1945, the Ursus factory was building a 45HP version of the Lanz Bulldog tractor under licence, before switching their allegiance to Zetor and then Massey Ferguson.

Another early example of technology transfer came in 1912 when the British-based agricultural steam engine manufacturer Clayton and Shuttleworth established a new company in Hungary. At that time, the farming industry in Hungary was still based on large semi-feudal estates, and these provided an important export market for steam engines built by British companies. The importance of the market encouraged Clayton and Shuttleworth to establish a joint venture with a Hungarian company, and the partners it chose were Hofherr Matyas and Schrantz Janos, who already owned a large machinery manufacturing company in Budapest. The new Anglo-Hungarian enterprise's name was Hofherr-Schrantz-Clayton-Shuttleworth, which, fortunately, was abbreviated to HSCS.

Production started with Hungarian-built versions of the British company's steam engines and threshing equipment, but petrol engines were added to the HSCS product range in about 1920, and the first petrol-powered tractor followed in 1923. Semi-diesel engines were available three years later. These were initially used as stationary or portable power units, but HSCS also used the same type of engine when it expanded into tractor production, and the company was one of the leading manufacturers of semi-diesels in both wheeled and track-laying versions from the late 1920s.

Tractor production came to an end when Hungary was occupied during World War II, but a small number of semi-diesel tractors were built after 1945 until the factory was reorganized under the new communist regime. Inevitably, perhaps, the original Anglo-Hungarian company name was too closely linked with old-fashioned British capitalism and was soon out of favour with the communist management. In 1951, the name was changed to the more politically correct Red Star Tractor Works, and, in 1960, the brand name on the tractors was changed to Dutra. This, apparently, is an abbreviation of 'dumper' and 'tractor', and the name

This HSCS R30-35 tractor powered by a single-cylinder, semi-diesel engine was built at the Anglo-Hungarian Hofherr-Schrantz-Clayton-Shuttleworth factory in 1935 and was exported to Australia. In 1990, it was shipped to England for restoration by the Shuttleworth Trust.

Dutra tractors from Hungary provided one of Europe's first 100HP models with four-wheel drive, and they were popular on big farms in both Eastern and Western Europe. The four-wheel drive system used big-diameter wheels at the front and rear, and the rigid frame ensured a small steering angle and a big turning circle.

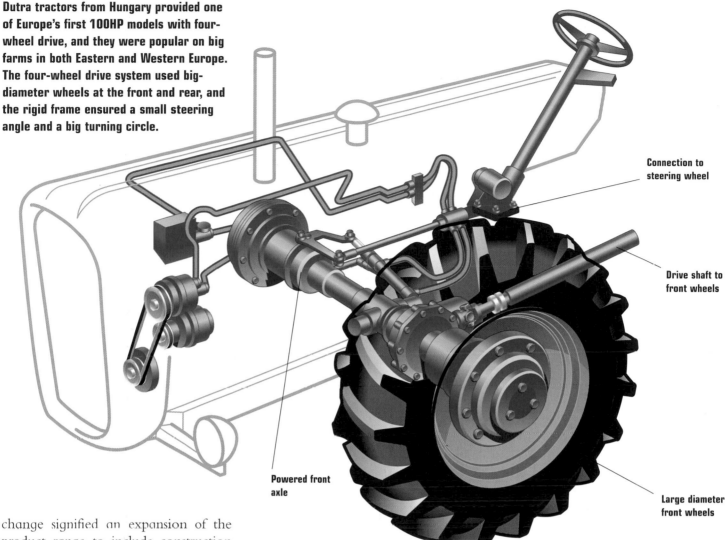

Connection to steering wheel

Drive shaft to front wheels

Powered front axle

Large diameter front wheels

change signified an expansion of the product range to include construction equipment as well as tractors.

As well as changing the company name, the new management also replaced the old semi-diesel engines with more up-to-date, multi-cylinder diesels, and a new range of tractors arrived in the late 1950s featuring four-wheel drive with equal-sized wheels and power outputs up to 100HP. The design included a number of advanced features such as front and rear wheel braking, and the Hungarian-built, six-cylinder Csepel DT613.15 series engine on the 100HP D4K-B model could be replaced by a more familiar Perkins power unit in tractors built for some of Dutra's export markets. The principle disadvantage of the Dutra design was the large turning circle required for the big front wheels and the rigid frame design, but the benefits of articulated steering, which would have overcome this problem, were not widely available in the late 1950s.

Dutra was one of the first European manufacturers to offer the combination of 100HP and four-wheel drive. Although the tractors were aimed mainly at the big Hungarian farms, which had begun to operate as collectives, they also achieved export success in other Eastern European countries, as well as in the West.

By the mid-1970s, there was a change of management when the former HSCS factory was taken over by the Hungarian Railway Wagon and Machine Company. The new owners became involved in another East–West collaboration when they signed an agreement to build high-horsepower Steiger tractors under licence. They were sold under the Raba Steiger brand name and were designed to meet the continuing demand for big tractors in

Hungary. For the managers of the big collective farms, they were a popular alternative to the big, high-horsepower tractors from the Soviet Union.

RUSSIA IN THE 1970S

Russia's big tractors in the 1970s included the 165HP Belarus 1500, powered by a V6 engine plus a donkey engine for starting. A new range of high-horsepower tractors has been available from the Kirov factory since the early 1990s, marketed in the West as the Peter the Great series, with a more up-to-date specification including articulated steering and a sloping bonnet line for improved forward visibility. The Peter the Great is available in 250HP and 350HP versions powered by six-cylinder turbo engines developing their rated power at 1900rpm. The tractor's

Allis-Chalmers built a series of tractors at its British factories for about 20 years from 1947. Production ended with the ED40 model, which was built at the company's Essendine factory in Lincolnshire between 1960 and 1968. It was powered by a 37HP diesel engine supplied by the Standard car company.

four-cylinder engine with 83mm (3.25in) bore and 89mm (3.5in) stroke. Production later moved to Stamford, Lincolnshire, England, where a Perkins P3 diesel engine was offered as an option for the Model B from 1954.

Model B production ended in 1955 when the new D270 model arrived, and two years later this was replaced by an improved version known as the D272. Engine options for the D272 included a paraffin-powered model equipped with a 26HP Allis-Chalmers engine and a 31HP diesel powered by a P3 engine. A special feature of the D270 and subsequent models was the live power

take-off operated by a hand lever. It allowed the operator to disconnect the drive to the rear wheels without stopping the power take-off shaft. A live power take-off was particularly useful for operating the Roto-Baler, the highly successful machine introduced by Allis-Chalmers in the late 1940s for making small, round bales, as it halved the number of clutch pedal movements needed to make each bale.

Allis-Chalmers' British production run ended with the ED-40 tractor, which arrived in 1960 and was available until 1968. The first production model was powered by a 37HP diesel engine built by the Standard car company, but this was stepped up to 41HP when an updated version with an improved hydraulic system arrived in 1963.

Minneapolis Moline appointed a British distributor for their UDS tractors, which were shipped from the American

factory without an engine. The distributor fitted British-built diesel engines, and customers were offered the choice of either a 65Hp Meadows power unit or a 46HP Dorman. With the Meadows engine fitted, the UDS was claimed to be one of the most powerful wheeled tractors available in Europe.

INTERNATIONAL HARVESTER

The biggest Anglo-American success story in the late 1940s began in 1949 when the first British-built International Harvester tractor rolled off the production line at the new International Harvester factory in Doncaster,

The first Massey-Harris tractors built in Britain were based on the 44k model, with the original vaporizing oil engine replaced by a Perkins P6 diesel. Production of the 744PD began in 1948. The '7' in its model name stands for British build and 'PD' was added for 'Perkins Diesel'.

International Harvester started building its British factory at Doncaster, Yorkshire, in 1939, but wartime supply problems delayed production until 10 years later. The first British-built IH tractor was the BM, based on the popular Farmall Model M and powered by an American-built engine.

Yorkshire. International Harvester was already well established in Britain, where imports of the smaller American-built Titan and Mogul models were among the most popular tractors on British farms during World War I. Between the wars, International Harvester continued to export large quantities of farm machinery to the British market, as well as wheeled and crawler tractors. Some of the tractors were shipped in kit form for assembly in Britain, and the company built its new Doncaster, England, factory in 1939, but tractor production had to be delayed until after World War II ended in 1945.

Manufacturing started in 1946 with a wide range of machinery while the factory was being equipped for tractor production to start in 1949. To stress the importance of the 1949 event, Tom Williams, then British Minister of Agriculture, drove the first British-built International Harvester tractor off the production line. It was a Farmall BM, based on the popular American-designed M series tractor, but with a 'B'

added to the 'M' to identify its British origins. The number-one BM tractor driven by the minister has survived and still belongs to descendants of the original owner.

A diesel-powered version of the BM, known as the BMD, followed in 1951, and two years later the BM and BMD, both based on a 1939 design, were

replaced by the new improved versions called the Farmall Super BM and Super BMD. The power output for both was increased, with the paraffin-powered version of the Super BM and the diesel engine in the Super BMD both delivering 50.5HP. A refinement on both models was the improved hydraulic system, with its control to alter the lift and lower rate for mounted implements; the hydraulic system could also be used to adjust the height of the pick-up hitch in order to simplify attachment of trailed equipment.

Following the introduction of the new Super models in 1953, International Harvester started planning the development of the new B250 tractor. It was announced in 1955 at Smithfield Show in London and was the first International Harvester tractor to be both designed and built in Britain. It was also the first tractor to be built at a separate factory in Bradford, England, that International Harvester had taken over and refurbished. The extra factory was needed to keep pace with the increased production schedules and had previously been the home of the Jowett range of cars and light commercial vehicles. The standard specification of the B250 included a 2363cc four-cylinder diesel engine producing 28HP rated output, and the gearbox provided five forward speeds.

The International Harvester success story continued as production expanded with the addition of crawler models, starting with the BT6 and BTD6 tractors available from 1953. The model range expanded during the 1960s to include the 14.2-tonne (14-ton) BTD20 track-layer, the biggest International Harvester tractor, built at Doncaster, England.

Tenneco, which already produced the Case and David Brown ranges, made a successful bid to buy the International Harvester tractor and farm equipment operations in 1985, and Case tractors gradually replaced the International Harvester models on the Doncaster production lines.

MASSEY-HARRIS

Massey-Harris was also part of the postwar trend for US-based tractor companies to expand into Britain. Exports to Britain had started in the nineteenth century, long before the Massey and Harris companies had joined forces. Massey-Harris later became one of the leading suppliers of machinery and, following its takeover of the J. I. Case Plow Co. in 1928, it also captured a slice of Britain's tractor market. British production of Massey-Harris equipment started at a factory in Manchester, England, in 1946. Implements and hay equipment were the first products, but tractors followed in 1948, starting with the 46HP MH 744PD. This was basically the Canadian-designed 44 tractor, with the number seven added to indicate British manufacture, while the addition of the letters 'PD' showed the engine was a Perkins diesel. The engine chosen for the new tractor was the highly successful six-cylinder P6, selected by the Massey-Harris engineers after a detailed evaluation of all the available diesel power options.

Tractor production was transferred from Manchester to a new factory in Kilmarnock, Scotland, in 1949. At this

The Singer car company moved into the tractor market for the first time in 1953 when it took over the Oak Tree Appliances company, manufacturers of the little OTA tractor. Singer merely changed the OTA's paint colour and updated the rear linkage.

stage, the 'P' was dropped from the model identification, and production of the 744D increased to about 50 tractors per week. The replacement for the 744D arrived in 1954, powered by a Perkins four-cylinder L4 diesel engine with 45HP rated output. The new version was the MH 745, but by this time Massey-Harris and Harry Ferguson's Coventry-based Ferguson tractor company had amalgamated, and production of the 745 ended in 1958.

THE LIGHTWEIGHT NEWCOMERS

Ferguson was one of the long list of new British companies that moved into tractor production in the years following the end of the war in 1945. Most of the newcomers were apparently expecting a big surge in the demand for small, lightweight models, and some companies rushed into production with tractors that were often poorly designed and underpowered. Examples include the AN3 and AN4 models from Newman Industries of Grantham, Lincolnshire. The power units for these were Coventry Victor aircooled petrol engines with twin horizontally opposed cylinders. Engine capacities were 804cc and 905cc, with rated outputs of 11HP and 12HP, respectively, and the weight of the tractors was about 750kg (1,650lb).

The Powersteer tractor announced in 1949 is also in this category. A Ford industrial engine powered this model with 10HP output according to the old RAC system in Britain. It was based on a tricycle design with a single wheel at the front. The front wheel, too small to ride over anything except firm ground, was designed to swivel with a caster action, and the steering was controlled by levers and operated through the rear wheels. Maxim Engineering, the London-based manufacturer, said the steering mechanism was based on the LG1 system, which apparently operated by disconnecting the drive to one of the rear wheels to give a swivelling action from the other wheel. Powersteer tractors were not a success.

Equally unconventional was the Bean Row-Crop tractor, designed by a Mr Bean and built from about 1946 by the Humberside Agricultural Products

company of Brough, East Yorkshire. The Bean tractor was designed specifically for working in row crops such as vegetables, and the 8HP Ford engine was probably adequate for the generally light tasks it would handle. The tractor was based on a rectangular steel girder frame mounted on a tricycle wheel arrangement, with the engine mounted near the back of the tractor. The driver sat just in front of the engine, with a virtually uninterrupted view of the hoes and other front and mid-mounted equipment carried on the frame. The Bean won a silver medal at the 1947 Royal Show and was available in various versions for about 20 years, with the last production version built by a Scotland-based engineering company.

Nuffield tractors were among the early converts to diesel power, using Perkins engines initially. The four-cylinder engine on this Universal Four tractor was available from 1954; it was made by British Motor Corporation or BMC, the car and truck group which then owned the tractor business.

Another of the lightweight newcomers was the OTA tractor made near Coventry, England, by Oak Tree Appliances. This was another example of a tractor powered by a small Ford engine and based on a steel girder frame carried on a tricycle layout. The Ford power unit in this case was a 10HP industrial version of the side-valve engine designed for a Ford car, and the power was delivered through a three-speed car gearbox. The single front wheel was steered by cables from the steering wheel, and the main frame of the tractor was raised to allow plenty of clearance for mid-mounted equipment. The original version of the OTA made its first appearance at the 1949 Smithfield Show, and a four-wheel model followed it in 1951.

Although the OTA was one of the best designed of the late 1940s batch of small tractors, sales were disappointing, and, by 1953, the manufacturers probably welcomed a takeover offer. The new owner was Singer Motors, one of the smaller British car manufacturers, which wanted to diversify into tractor production. Choosing OTA as its takeover target was an odd decision, partly because the OTA tractor was obviously not a success and needed major design changes and partly because it was designed to use major components supplied by Ford, one of Singer's strongest rivals. Presumably Singer had plans to invest money in design improvements for its new tractor; however, apart from a name change to the Singer Monarch, a new colour scheme and a redesigned rear linkage, the tractor remained basically unaltered. Production ended in about 1956, when Singer was taken over by one of its bigger competitors, the Rootes Group.

MORRIS MOVES INTO TRACTORS

Some of the companies moving into the tractor market in the years following 1945 did their market research more thoroughly and came up with a more accurate idea of the type of tractor the market would demand, and Morris Motors was one company that came up with the right product. Morris was one of the biggest British manufacturers of cars and commercial vehicles, and the success of its move into tractor production may have encouraged Singer to follow its example.

Morris Motors was established by William Morris, who later became Lord Nuffield, and Nuffield was the brand name chosen for the new tractor announced in 1948. It was called the Nuffield Universal, and it was available as an M4 version with standard wheel

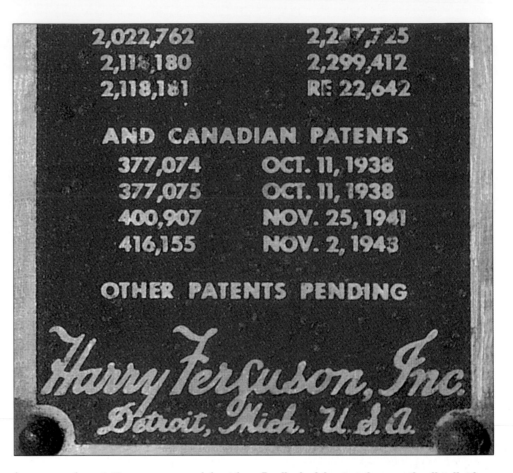

layout and an M3 rowcrop model with a tricycle layout designed mainly for the American market. The power unit was the four-cylinder, side-valve engine that had powered thousands of Morris Commercial trucks supplied to the British Army during the war, and it was linked to a five-speed gearbox. Power output was 41HP for the petrol version and 38HP when adapted to burn paraffin. A diesel version equipped with a 48HP Perkins P4 engine arrived in 1950.

When Morris Motors and its associated companies including the MG, Riley and Wolseley car businesses plus the Nuffield tractor operation became part of the British Motor Corporation (BMC) in 1952, the Perkins engine was replaced by a 45HP BMC diesel. In 1962, there was a production move from a former Wolseley car factory in Birmingham, England, to a new plant at Bathgate, Scotland. Meanwhile, the tractors were earning an excellent reputation for performance and durability, and the international marketing strength of the parent company helped to build up substantial export markets for Nuffield in North America, Australia and parts of Africa.

Ford's decision to take over the distribution of the new 8N tractor left Harry Ferguson with a big marketing organization in North America, but no tractor to sell. To solve the problem, he formed Harry Ferguson, Inc., to build TO series Ferguson tractors and sell them in direct competition with the Ford 8N.

Further product development included a power boost to 60HP and a new transmission with 10 speeds for the appropriately named 10/60 model; in 1965, the Mini tractor was added to the range. This was an imaginative attempt to produce one of the first genuine compact tractors designed for small farms, horticulture and amenity work. A 15HP engine supplied the power, but, when this proved to be inadequate, it was replaced by a 25HP direct injection diesel engine.

Meanwhile, BMC faced serious problems, including disruption caused by frequent strikes, plus a reputation for poor reliability. The Nuffield tractor business – by now the poor relation – was starved of investment, and this became increasingly obvious as the product development failed to keep pace with competitors. This included the

The Detroit-built Ferguson TO-20 tractor was based on the design of the TE-20 model from the Ferguson factory in Coventry, England, and most of the differences were due to sourcing some components from American instead of British suppliers. Even the engines were the same.

Mini tractor, which was soon losing ground to new competitors with better specifications from Kubota, Iseki and other Japanese companies.

In 1969, a name change from Nuffield to Leyland, together with new styling and a two-tone blue colour scheme instead of Nuffield's familiar poppy red, helped to revive interest in the tractors, in spite of the somewhat modest mechanical improvements.

Another colour scheme change, this time to Harvest Gold, came in 1980, and this was followed in 1982 by a change of ownership and yet another name change.

The ownership change came when the Leyland group sold off some of its non-car and truck businesses, and the name change was a revival of the old and much respected Marshall name. The tractors were transported from the Leyland plant in Scotland to the old Marshall factory at Gainsborough, Lincolnshire, for the pre-delivery inspection and distribution to dealers. Finance for the major design changes that were essential was still not available, and demand continued to shrink. Various attempts to keep the

company afloat included importing Steyr agricultural tractors from Austria and a range of compacts from Italy and rebadging them as Marshalls. By the early 1990s, however, the most important part of the business was supplying parts for the large numbers of Nuffield and Leyland tractors found throughout the world.

FERGUSON'S POSTWAR SUCCESS

The biggest success story among the postwar newcomers was the company set up by Harry Ferguson. Ferguson had assumed that Ford would eventually replace the increasingly outdated Model N with a British-built version of the 9N or its replacement. The Ford company had other ideas, however, and the partnership between Ford and Ferguson ended in an acrimonious and costly lawsuit (see chapter 3). Instead of adopting a Ferguson System tractor and consulting Harry Ferguson on its plans, the Ford management at Dagenham announced the new E27N tractor. The new model was based on the Model N and its Model F ancestor, but had new styling and a number of mechanical design improvements that included pensioning off the outdated wormwheel final drive of the F and N series tractors.

E27N production began in 1945, and a Perkins-powered diesel version followed in 1948. A hydraulically operated linkage was added later without infringing Ferguson patents. In spite of its success, with production peaking at more than 50,000 in 1948, the E27N was merely a stopgap to allow time for the company to develop the all-new Fordson Major series E1ADKN. The New Major, as it was called, was announced at the 1951 Smithfield Show and the most interesting feature was the new engine. It had been under development since 1944 and was based on a four-cylinder block with 3.6 litres (220 cu. in) capacity. The New Major was produced in petrol, paraffin and diesel versions.

It was increasingly obvious that Harry Ferguson would have to make his own tractor production arrangements, and he did so through another partnership deal. The fact that both his earlier partnerships – with David Brown and Henry Ford – had ended in friction did not discourage Ferguson from entering into a similar arrangement with Sir John Black's Standard Motor Co. Standard occupied the Banner Lane factory in Coventry, England, used during World War II to build engines for military aircraft. There was plenty of spare space in the factory to build tractors as well as the Standard range of cars and engines, and Sir John's company was responsible for production, while Ferguson looked after design and marketing.

The new tractor was the Ferguson TE-20 series, and production started in 1946. The letters stood for 'Tractor England', the styling closely resembled

FERGUSON TE-20

Manufacturer: Ferguson
Model: TE-20
Production started: 1946
Power unit: First two years:
Continental overhead-valve
engine with 81 x 95mm (3.19 x
3.74in) bore and stroke
Power output: 23HP
Transmission: Gearbox with four
forward speeds and one reverse

the Ford Model N, the paint finish was Ferguson's favourite shade of grey and the power unit during the first two years of production was a 1966cc US-built Continental Z-120 producing 23HP. In 1948, the Continental engine was replaced by a new four-cylinder engine designed and built by the Standard Motor Co. The capacity of the replacement engine was 1849cc, and the power output was 25HP at 2000rpm. The same engine was used to power Standard Vanguard cars and vans, providing substantial production economies.

A diesel-powered Ferguson arrived in 1951, powered by a new 2092cc engine built by Standard. Freeman Sanders, one of Britain's most experienced diesel engineers, had helped to design the Standard engine. He had been responsible for diesel engine development at Fowlers during the 1930s, and he was also responsible for the novel V-4 diesel engine that powered the Turner Yeoman of England tractor built in small numbers between 1949 and 1957.

As we saw in chapter 3, Ferguson tractor production at Banner Lane started just in time to supply tractors to Harry Ferguson's US distribution company after the Ford company's decision to set up its own marketing company for the new 8N tractor. TE-20s maintained the American sales operation while Harry Ferguson was arranging to build Ferguson tractors in the United States. The American plans were completed with extraordinary speed. In January 1948, the Ferguson company bought a plot of land in Detroit, building work on the new

factory began in February and, on 26 July of the same year, Harry Ferguson drove the first of the new TO-20 tractors off the production line. TO stood for 'Tractor Overseas', but both versions were equipped with the Continental engine and the main difference between the TE and TO models was that some TO components were sourced from American manufacturers.

A NEW DIRECTION

The 1953 announcement that Massey-Harris, the Canadian-based machinery giant, was merging with the Ferguson company came as a surprise to many people in the farm equipment industry, and it is no less surprising now. Although

the deal was described as a merger, it was actually a takeover, with the Massey-Harris interests retaining full control, and it is difficult to understand why Harry Ferguson was so willing to sell his company. His tractors were achieving worldwide success, his company was making substantial profits and he was still full of ideas and future plans.

One possibility is that Ferguson assumed he would play a leading role in the enlarged Massey-Harris-Ferguson company. Under the terms of the agreement with Massey-Harris, Harry Ferguson was given full control over tractor development, and he may have seen this as an opportunity to make his ideas more widely available to farmers throughout the world. If this was the strategy, Ferguson was to be sorely disappointed, as he was never given the full control he had expected, and this later caused his resignation from the company. Another explanation given for the sale is that Ferguson had some concerns about his health and saw the Massey-Harris deal as an opportunity to secure the long-term future of the company he had built.

Having sold his company, Harry Ferguson set up his own research organization, and one of the projects was a four-wheel drive system to improve the stability and safety of cars. This was somewhat surprising, as Ferguson had never shown much enthusiasm for four-

Harry Ferguson and his engineers made a major contribution to the Ford 9N tractor's design; its three-point linkage and hydraulic controls were based almost entirely on Ferguson's patents. This is probably why he decided to base his TE-20 and TO-20 models on the highly successful Ford tractor.

wheel drive tractors. His views on using four-wheel drive for tractors were shared by many in the farm equipment industry at that time, but opinions would soon be changing. Many farmers, particularly those with heavy soils, wanted improved pulling efficiency from their tractors, and the only way to achieve this was by using tracklayers. This was almost 50 years before the development of rubber

TURBOCHARGING

Turbocharging was introduced originally as a way to extract more power from diesel engines, with increases typically in the 20 to 25 per cent range.

Turbos are efficient because they use waste energy from the engine's exhaust. The rapidly moving stream of gases from the combustion chamber spins a tiny turbine at speeds of 100,000rpm or more, and this drives an impeller forcing extra air through the inlet manifold. Oxygen in the air from the impeller allows more fuel to be burned in the combustion chamber and this produces the extra power output.

As well as producing more power, turbochargers also have other benefits. They tend to reduce the engine noise level, particularly from the exhaust, and they also improve the efficiency of the combustion process. More efficient combustion helps to achieve more work from each litre of fuel, and it also produces cleaner exhaust emissions, and this is why most tractor engines above about 85HP now have a turbo.

tracks (see chapter 7) and the old-fashioned steel tracks had a number of serious disadvantages.

Crawler tracks are the most efficient way to turn engine power into drawbar pull, particularly on wet soil where the performance of a conventional two-wheel drive tractor is likely to be reduced by wheelslip. A number of independent tests have shown that putting the power through all four wheels can improve traction by about 10

Joining two tractors together to form a single unit with double the power was a good idea in theory, but the problem was designing a link strong and flexible enough to allow a tight turning circle. The problem was solved by using a turntable and two rams to steer the front unit hydraulically.

per cent in conditions where wheelslip is likely to be a problem, with the biggest performance increase coming from tractors with large-diameter wheels at the front and rear. Four-wheel drive with smaller front wheels – sometimes called front-wheel assist – is less effective, but still significantly better than a basic two-wheel drive arrangement.

FOUR-WHEEL DRIVE PIONEERS

Harry Ferguson's viewpoint on four-wheel drive tractors may have been influenced by the disappointing results achieved by companies that had already tried the idea. Four-wheel drive was

certainly not new. It had been available as early as the 1880s on steam traction engines, and several tractor manufacturers had offered four-wheel drive with limited success. The Four-Wheel Pull model built by the Olmstead company of Great Falls, Montana, was available from about 1912, with the engine power delivered through equal-diameter front and rear wheels. The idea attracted little interest, and this lack of interest was repeated when Samson Iron Works of Stockton, California, began building its four-wheel drive Iron Horse tractor in 1919.

One of the problems with large-diameter front wheels is that there is not enough space to achieve a reasonably large steering angle, and this affects the manoeuvrability of the tractor for headland turns and working where space is restricted. Using articulated or bend-in-the-middle steering with the tractor split into front and rear sections that are linked by a pivot point can solve this problem. This is now a standard design feature for the biggest four-wheel drive tractors, but it was regarded as little more than a novelty when first used in the 1920s. Lanz offered pivot steering on the four-wheel drive Allrad version of its Bulldog tractor in 1923; two or three years later the Italian Pavesi company produced the P4 tractor with four-wheel drive and pivot steering; however, sales of both tractors were disappointing.

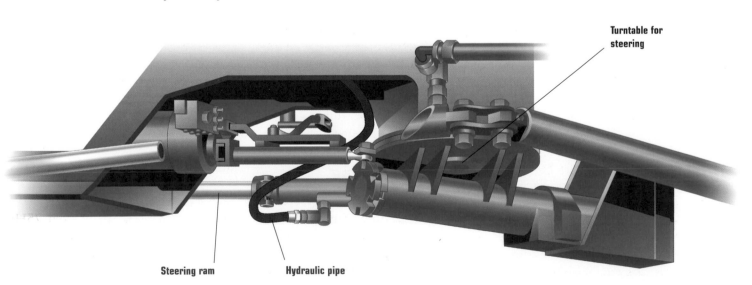

Turntable for steering

Steering ram   Hydraulic pipe

Since the 1970s, pivot steering has become established as the standard system for high-horsepower four-wheel drive tractors, and it is also used at the other end of the size scale on some compact tractors used for amenity work. This is a particular speciality of some of the Italian manufacturers of compacts, including Antonio Carraro, Pasquali and some Goldoni models, but it was also used as long ago as the mid-1950s on a pivot-steer horticultural tractor made by the Howard Rotavator Co.

Massey-Harris achieved more success with its General Purpose tractor. Production began in 1930, making it the first tractor designed and built by Massey-Harris. The General Purpose was given a 15HP rating at the drawbar and 22HP at the belt pulley, with a four-cylinder Hercules engine providing the power. The power was delivered through four large-diameter wheels, and the tractor's rigid frame layout and front-wheel steering resulted in poor manoeuvrability. The fact that the General Purpose tractor achieved a modest level of success probably reflects the marketing strength of Massey-Harris, rather than any upsurge in demand for four-wheel drive.

The sales breakthrough for tractors with power delivered through the front and rear wheels came not from the big

manufacturers such as Massey-Harris, but from small-scale specialist companies in the United States and Britain.

BREAKTHROUGH SUCCESS

By the late 1950s, there were six companies in Britain either building four-wheel drive tractors or producing four-wheel drive conversions of standard models. The list of specialist four-wheel drive companies included Bray, with conversions based on the Nuffield range; County and Roadless, which used Ford and Fordson skid units to produce their conversions; and Matbro and Muir Hill, which built a series of tractors with four-wheel drive through equal-diameter wheels.

Also on the list was the Ernest Doe company, a Fordson tractor dealer based in Essex, England, and it based its Triple-D tractors on an idea developed by George Pryor, one of its farmer customers. To increase efficiency on his heavy clay soil, Pryor wanted the extra traction offered by four-wheel drive as well as increased horsepower, and this was achieved by linking two of the new Fordson Major tractors together. He removed the front wheels and axles from both tractors and joined the two together by using a steel turntable to provide a pivot point for the steering. A driver on the rear Fordson controlled

One of the German tractor manufacturers which helped to establish systems tractor is Fendt. This Fendt Xylon shows all the basic features of a systems tractor, including the mid-mounted cab position, load space behind the cab and provision for mounting p-t-o powered equipment front and rear.

both tractor engines and transmissions, and the complete unit was steered by operating a pair of hydraulic rams to swivel the front section to the left or right. In spite of its 5.79m (19ft) total length and its full-diameter front and rear wheels, the Doe tractor's pivot steering provided reasonably nimble steering. The two Fordson engines produced just over 100HP, more than any other wheeled tractor available in Britain at that time, and the four-wheel drive traction helped to increase output in difficult soil conditions.

Ernest Doe built a production version of the tractor in 1958. It was based on the Fordson Power Major and was the first of more than 300 tractors built before production ended in 1966. Later versions included the Doe 130, based on two Ford 5000 power units totalling 130HP, and the final model was the 150HP Doe 150 based on the uprated 75HP Ford 5000. Several other dual power unit tractors were built by farmers and dealers in Australia, France and the

This sectioned diagram of a John Deere 70 Series tractor with articulated steering shows how the drive shafts delivering power to the rear wheels cope with the hinge action of the steering mechanism. Deere 70 Series tractors have power outputs in the 250 to 400HP range, using six-cylinder engines equipped with a turbo and intercooler.

United States, but the idea lost its appeal as the big tractor companies developed bigger tractors with four-wheel drive, avoiding the extra cost and maintenance needs of two separate power units.

The biggest and most successful of the British four-wheel drive conversion specialists was County Commercial Cars, established by the Tapp brothers in 1929 to build special vehicles based on the Ford truck chassis. Tractor conversions began in 1948 with a tracklayer based on the Fordson E27N; by the end of the year, 50 of them had been supplied to the Ministry of Agriculture. It was called the County Full Track or CFT, and it was the first of a long series of tracklayer conversions based on Ford skid units.

Four-wheel drive conversions started in 1954 when the first County Four-Drive was built. It was based on the Fordson New Major, was designed with equal-diameter front and rear wheels, and used a chain and sprockets to transfer the drive from the rear to the front axles. Steering was based on a skid system borrowed from its tracklayers and operated by a pair of levers, and the tractor was designed specifically to work in the West Indies' sugar cane fields.

The number of County tracklayer conversions reduced steadily over the

Right: Steiger's system for combining four-wheel drive with an articulated steering system is shown in this diagram. Extra flexibility provided by the universal joints at both ends of the two drive shafts reduces the angle in the drive line as the steering system operates. This means less stress in the drive line during turns.

Drive shafts

Pivot point
for steering

next 30 years or so and came to a halt in 1965, while the production of four-wheel drive tractors increased. There was a wide range of models, nearly all of them based on Ford tractors and equipped with large-diameter front and rear wheels. In spite of the big front wheels, all the tractors had rigid frames, which earned them a reputation for poor manoeuvrability, but they provided effective traction and were popular in a long list of export markets as well as in the United Kingdom.

County's range included a number of unconventional tractors, among them

the Sea Horse, an amphibious version of the County Super-4 based on the Fordson Super Major. Modifications needed to produce the Sea Horse included steel flotation tanks front and rear, plus additional buoyancy chambers built into the wheels. All electrical equipment was removed from the engine, and an inertia starter was fitted instead, and the engine and transmission were waterproofed. The Sea Horse was designed to work in extremely wet conditions, but to demonstrate its capabilities County drove a Sea Horse 45km (28 miles) across the English Channel from France to England, earning the tractor a place in the *Guinness Book of Records*.

Of more practical interest were the various models in the County FC or Forward Control series. As the name suggests, the layout of these tractors

placed the cab right at the front of the tractor, leaving a large load space at the rear. The space gave the County FC tractors a load-carrying ability similar to the transport tractors described in chapter 5, but the heavyweight County with its awkward steering was more suitable for moving heavy equipment short distances on a construction site than for general transport duties.

THE BIG COMPANIES TAKE OVER

The obvious threat to four-wheel drive specialists such as County, Doe and Roadless was that, once they had done all the hard work and established the demand, the big companies in the tractor industry would develop their own four-wheel drive models and take over the market. This is exactly what happened, as companies such as Ford and Massey Ferguson introduced their own four-wheel drives, usually – but not always – with front-wheel assist.

Four-wheel drive is not the only arrangement for boosting traction. The Versatile Big Roy tractor, described in chapter 5, was equipped with eight-wheel drive, and some of the current models in the Horsch range of high-horsepower tractors from Germany put their engine power through three wheels. Another unconventional drive arrangement arrived in 1975 on the Valmet 1502 model from Finland. Valmet, which changed its name to Valtra in 2001, designed the 136HP tractor with six-wheel drive, replacing the usual pair of rear wheels with a bogie unit carrying four wheels. The makers claimed that vertical movements as the bogie travelled over rough ground improved the wheel grip and gave the driver a smoother ride.

In the North American tractor market, much of the postwar development of four-wheel drive was in the hands of specialist companies, just as it was in Britain. The US and Canadian specialists

The Valmet tractor range from Finland, now sold under the Valtra brand name, introduced reverse drive or bi-directional tractors in the 1980s. This 83HP Valmet 705 tractor has the seat turned through 180 degrees to allow the driver to operate the controls while facing to the rear, using a second, smaller-diameter steering wheel.

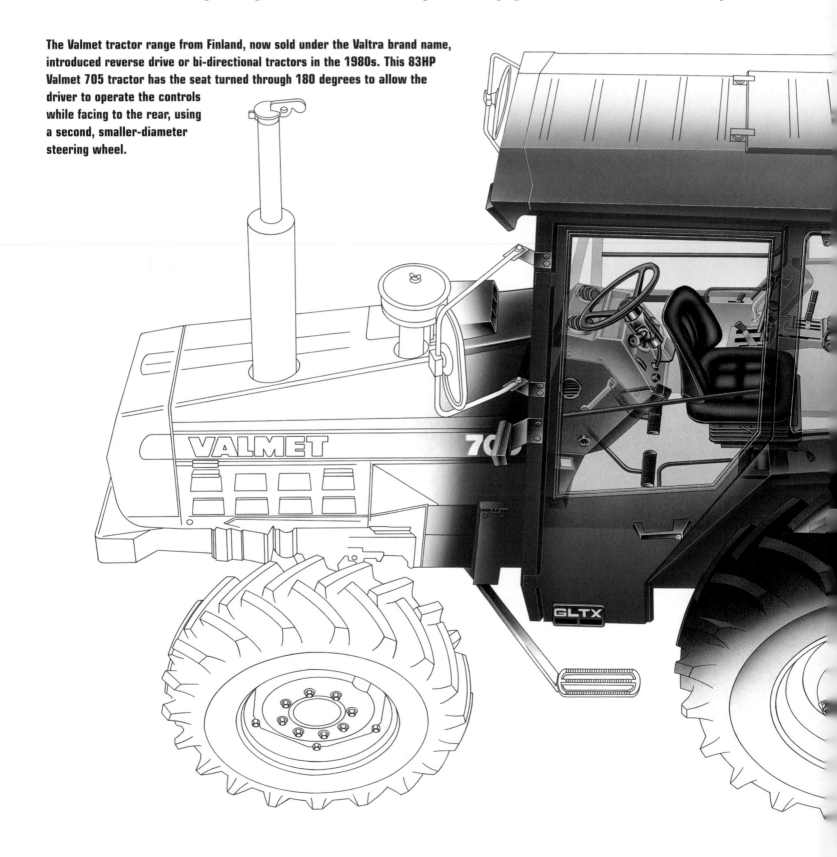

were less exposed to competition from the established tractor companies, however, as they concentrated on a different sector of the market. The four-wheel drive leaders were Versatile in Canada and the Steiger company based at Fargo, North Dakota. The Big Bud company of Havre, Montana, was also prominent. Unlike the equivalent

companies in Britain, however, the North American specialists were all concentrating on the high-horsepower end of the market where they were less exposed to direct competition from the big established manufacturers.

THE HIGH-HORSEPOWER MARKET

High-horsepower tractors have a long tradition on big-acreage farms in the United States and Canada, starting in the early days of the tractor industry.

One of the old heavyweights was the 30-60 tractor built by the Minneapolis Steel and Machinery Co., one of the parent companies of Minneapolis-Moline. The tractor was sold under its Twin City brand name and was available for about seven years from 1913. A technically advanced engine with six cylinders produced the 60HP rated output at a time when many of the big tractor companies were still using two cylinders.. The weakness of the Twin City 30-60 was its 12.7-tonne (12.5-tonne) weight, giving what we would now regard as a poor power-to-weight ratio. The new generation of high-horsepower tractors arriving in the 1950s was different. They were certainly heavy, but they offered much more horsepower, plus the extra traction efficiency of four-wheel drive through equal-sized wheels, and pivot steering ensured good manoeuvrability.

The Steiger company had its origins in 1957 when the Steiger brothers decided they needed a bigger tractor with more pulling power to cope with the work-load on their 1600-hectare (4000-acre) farm near Red Lake Falls, Minnesota.

The big tractors available at that time did not meet Douglass and Maurice Steiger's requirements, so they adopted the do-it-yourself approach and built a tractor to their own design in the farm workshop. They fitted a 130HP engine and designed the tractor with pivot steering and four-wheel drive, and the finished tractor completed more than 10,000 working hours on their farm.

Neighbouring farmers heard about the home-built tractor and its performance, and the Steigers were soon being asked to build similar tractors for other farmers in the area. The business quickly outgrew the farm workshop, and extra help was hired to cope with the demand. At this stage, there were three models, all powered by Detroit diesel engines delivering from 196HP to 328HP rated output. In 1969, the Steiger brothers and local farmers and businessmen formed a new company, and production was moved to a new factory at Fargo, North Dakota. Demand continued to increase, with sales totalling $20 million in 1973 and rising to $104 million in 1976.

Much of the expansion was due to the success of the Series II tractors introduced in 1974. The 1974 models included the Tiger II weighing 13.5 tonnes (13.4 tons), including ballast weights and powered by a V-8 Cummins VT-903 turbo engine which produced 262HP at the drawbar. The top model in the Series III available from 1976 was the Panther III ST-325, which had 270HP at the drawbar when tested at Nebraska.

Instead of competing with the products from most of the established tractor makers, the Steiger models were complementary. A substantial part of the Fargo factory's output was building big four-wheel drive tractors for Allis-Chalmers, Ford and International Harvester. In 1986, the company was bought by Case IH to continue building high-horsepower tractors for the Case range. Tractor production at the new Versatile factory in Winnipeg, Manitoba, started in 1966 with the four-wheel drive D-100 and G-100 models. They were powered by Ford and Chrysler engines, respectively, and they were the

VALMET OY 705

Manufacturer: Valmet Oy
Model: 705 with reverse drive
Production started: 1984
Power unit: Four cylinders, turbocharged, 4.4 litres (269 cu. in)
Power output: 83HP
Transmission: High/low ratio gearbox giving eight forward and four reverse gears

first tractors in an ongoing series that established Versatile as one of the leading four-wheel drive specialists. Some tractors from the Versatile factory were supplied to the Italian-based Fiat company to be sold through Fiat dealers; they later became part of the Ford New Holland tractor range after the company was bought by Ford in 1987.

One of the most innovative tractors produced by Versatile was the model 150 two-way or bi-directional tractor. Production of the 150 started in 1977 and bi-directional models in various versions have been part of the Versatile range ever since. The bi-directional's special feature is that it is designed for working forwards or backwards, with the driver's seat swivelling through 180 degrees to allow the driver to face whichever way the tractor is working. The steering wheel and major controls either swivel with the driver's seat or are duplicated at the rear of the cab, and there are attachment points for implements on the front as well as the rear.

The list of jobs which can be dealt with more efficiently when the tractor is working in reverse include materials handling with an industrial-type loader. Reversing the seat means the driver is looking through the rear window of the cab and is much closer to the loader, giving excellent visibility of the lifting and loading operations. Other jobs for a reverse-drive or bi-directional tractor include buckraking and operating some types of timber-handling equipment; in Europe, some types of harvesting machinery for forage and sugar beet are designed to work on the back of a bi-directional tractor working in reverse.

Versatile is firmly established as the leading manufacturer of tractors designed for two-way working, and most of the European bi-directional tractors have been cab conversions based on standard models from a long list of manufacturers including Fendt, Same, Renault, Mercedes-Benz, Massey Ferguson and Valmet. The Moffett MFT or Multi-Function Tractor was also a conversion, but in this case the conversion was more thorough.

The Moffett tractor was designed and built in Ireland, the first agricultural tractor to be produced in Ireland since the end of Fordson Model N production at Cork in 1932. Moffett production started in 1991, based on a 90HP Massey Ferguson skid unit, but this was replaced in 1995 by a new model with a 100HP New Holland engine and transmission. With the driving seat in the rear-facing position, the MFT could do the job of a heavy-duty loader, with the weight near the tractor's rear axle, where much of the tractor's strength is concentrated. Removing the loader attachment and switching the seat to the conventional forward-facing position allows the MFT to work as a conventional tractor for jobs such as ploughing. MFT sales totalled about 500 tractors before production ended in about 1998.

SYSTEMS TRACTORS

Systems tractors are another approach to altering the design of the standard tractor to improve working efficiency. Designed for maximum versatility, the design features of systems tractors include implement mounting points and power take-off drives at the front and the rear, and four-wheel drive through equal-sized wheels. The cab is normally moved to the middle or front of the tractor to allow space for a small load-carrying platform over the rear wheels. The idea originated in Germany during the 1970s, and the pioneers were Deutz-Fahr with its Intrac models and Mercedes-Benz with the MB-trac. Production of both tractors began in 1972, and they have had a major influence on subsequent tractor development.

VERSATILE 256

Manufacturer: Versatile Farm Equipment Co.
Model: 256 Bi-directional (later the Ford Bi-directional)
Production started: 1984
Power unit: Four-cylinder, 3.9 litres (239 cu. in) Cummins turbocharged engine
Power output: 100HP
Transmission: Three-speed hydrostatic

It was the Canadian-based Versatile company that produced the first tractor designed specifically for two-way or bi-directional operation. Its 150 model was available from 1977 and it was followed by further versions offering increased power.

Some of the features they helped to pioneer, including a three-point linkage for front mounting, are now available on standard as well as systems models.

One of the advantages of the system tractor's front and rear mounting is that it allows two or more jobs to be completed in a single pass, and an example would be a front-mounted rake to combine several swaths ready for the baler at the rear. An alternative option is to use the front linkage for equipment such as an inter-row hoe that works best on the front of the tractor where it is easily visible for maintaining accurate control. The rear loading platform can be used for carrying extra bags of seed or fertilizer,

or for mounting an extra spray tank to give more capacity and reduce the down time needed for tank refills.

Although the Intrac and the MB-trac that introduced the systems approach to tractor design are no longer available, other manufacturers have adopted all or some of the features that were available on these models. Fendt, also based in Germany, produced a systems tractor during much of the 1990s, and the JCB Fastrac is a systems tractor with the added advantage of faster travel speeds.

THE LANZ ALLDOG

Tractors designed to provide a large choice of mounting and attachment points for implements provide a different approach to versatility. One example is the Lanz Alldog tool-carrier tractor from Germany, available between 1951 and 1959. The Alldog was designed with the engine and the driver positioned at

the rear of a rectangular steel tube framework, leaving a large space in front on which equipment could be mounted or underslung. At one stage, the list of equipment suitable for the Alldog to carry or power extended to 50 items, and these included a tiny portable milking machine that could be taken out to the field where it was powered by the tractor engine. There were load boxes to mount on the framework, mowers, ploughs, cultivators and even a wrap-around combine harvester and a semi-mounted sugar beet harvester.

In theory it was a good idea, ensuring that one tractor could do all the work on a small mixed farm. The problem was that the Lanz engineers were not sufficiently generous with the engine power on the Alldog. The first production version was the A1205 model, which was powered by an inadequate 12HP petrol engine; with this amount of power available, the ploughing or sugar beet harvesting work rates must have been extremely slow. Presumably someone mentioned the lack of power to Lanz, and, when the updated diesel version arrived, it provided a one-horsepower power boost to 13HP. The air-cooled diesel engine

Germany's Heinrich Lanz company concentrated on versatility when it introduced the first of its Alldog tool-carrier tractors in 1951. The tubular steel main frame could be used to attach equipment at the front, middle and rear.

was an unusual dual-fuel design, equipped with a spark plug for starting on petrol before switching to diesel. The final version, called the A1806, was available from 1956 to 1959, and it was powered by an MWM diesel engine with liquid cooling and an 18HP output. Some of the A1806 Alldogs were

painted green and yellow to indicate that they were sold after the Lanz company had been bought by John Deere.

THE UNI-FARMER

The United States' leading candidate in the multi-function power unit market was the Uni-Farmer tractor produced initially by Minneapolis-Moline, but later available under the Uni-System name from New Idea. The original Minneapolis version was designed as the base unit for a series of specially designed harvesting machines. It was originally introduced in 1950 and was described in the company's publicity material as the 'greatest development' in the history of Minneapolis-Moline.

LANZ ALLDOG

Manufacturer: Heinrich Lanz
Model: Alldog A1315
Production started: 1954
Power unit: Air-cooled single cylinder diesel
Power output: 13HP
Transmission: Six forward gears and a reverse

Steiger was one of the first of the four-wheel drive specialists for the North American market. The first tractor the Steiger brothers built was for their farm in Minnesota, using a 238HP Detroit Diesel engine. Later models included this Panther ST325 model finished in Steiger green.

One feature the Uni-Farmer shared with the Lanz Alldog was lack of power, (the four-cylinder V-206 tractor engines only developed 38HP), and this helped to ensure that the work rate would lag behind most tractor-powered or self-propelled combine harvesters.

There were just two attachments when the Uni-Farmer was launched: a combine harvester with a 2.7m (9ft) cutting width and the Uni-Husker, a two-row corn picker and husker. The Uni-Forager, which was released in 1953, had rowcrop and pick-up headers for dealing with all types of silage. When New Idea took control of what was now called the Uni-System, it offered three different power outputs, including a more competitive 100HP-plus version. It also tried to add p-t-o powered tillage equipment to the range of attachments, but with limited success. Uni-System production ended in the early 1970s.

SHAPING THE MODERN TRACTOR

Driver comfort and safety, neglected for decades, are now important priorities. Safety cabs have been forced on the tractor industry by legislation which, coupled with cab suspension systems and sprung axles, has allowed faster, safer working speeds. Other major developments include the introduction of rubber tracks, which have dramatically improved the performance of crawler models, turbocharged engines and a new generation of easy-to-use CVT drive systems.

Although driver safety became an important issue during the 1960s, with legislation enforcing the use of safety cabs, it was not until the mid-1970s that the comfort and convenience of drivers moved any distance up the priority scale. For the first 80 years or so of tractor history, the only concession to comfort for the majority of drivers was a plain steel seat on a sprung mounting. In fact, even the seat was absent from some of the earliest tractors, and the driver was expected to stand as he operated the controls.

Weather protection was quite unusual. A few of the early tractors were equipped with a canopy-type roof offering some protection from the rain. On some models, however,

Production of JCB Fastrac tractors started in 1991, supplying high speeds for transport work, plus four-wheel drive and a unique self-levelling suspension to provide effective pulling power for ploughing and other field operations.

the canopy covered the engine to keep rain out of the primitive electrical and fuel systems, but it did not extend over the platform at the rear, where the driver remained exposed to the elements.

Nobody, it seems, was particularly concerned about the lack of comfort. Improving reliability and performance was the priority for the companies that designed and built the tractors. The farmers who bought the tractors – and on many farms, they also did some or all of the tractor driving – were generally unwilling to pay extra for a model with a better seat or a simple cab. Many of the drivers had previously walked all day behind a team of horses, and the idea of riding even the most Spartan of tractors probably seemed a big improvement.

There were, of course, some exceptions to the general rule about lack of comfort. For some reason, many crawler tractors, and particularly those made by Caterpillar, were using seats with padded cushions and backrests from the mid-1920s, although it is not clear why drivers of tracklayers were singled out for such special treatment. Springs have appeared from time to time, and one of the earliest examples was the prototype Ford tractor built in 1907 (see chapter 2), and they also featured occasionally on other tractors during the next 70 years or so. The Mercedes-Benz MB-trac, the first and the most successful of the systems tractors, achieved a technological breakthrough by providing a suspension system on a powered front axle.

The first really serious attempt to make tractor driving a more comfortable occupation arrived in 1938 when Minneapolis-Moline announced its new UDLX tractor.

MAKING COMFORT A PRIORITY

Minneapolis-Moline called its new model a 'Comfortractor', and in the company's publicity it was billed as 'The World's Greatest Tractor'. It was described as a three- or four-plough tractor, but the sales leaflet is surprisingly reticent about quoting a horsepower figure and the UDLX was not tested at Nebraska.

For well over half a century, the only concession to driver comfort on most tractors was a shaped metal seat mounted on a springy steel support. Tractor manufacturers and their customers share the blame for giving safety and comfort such a low priority.

The decision to build the UDLX followed a market research survey suggesting that more than 50 per cent of farmers questioned would like a tractor with a cab. The new tractor provided the cab and much more besides, but the market research proved to be misleading, as few of the farmers who wanted a tractor with a cab were actually willing to pay for it. Engineers at Minneapolis-Moline designed a steel cab with glazed windows and noise-absorbing insulation for the Comfortractor. The UDLX was the first tractor designed and built with an enclosed cab, although it was also available without a cab. Inside the cab, the equipment list was more generous than that of many 1930s cars. It included a radio and a heater, plus a wiper and a defroster for the front window. There was an electric clock, a roof light and an illuminated instrument panel, and the M-M UDLX was almost certainly the first farm tractor equipped with both a cigar lighter and a built-in ashtray.

With two upholstered seats in the cab and a top speed of 64km/h (40mph) on the road, the UDLX could spend the day ploughing and then provide suitable transport for a romantic trip to a restaurant or the movies in the evening. Safety items included full road lighting equipment front and rear, a bumper for protection at the front plus self-energizing Bendix brakes, and there was safety glass in all the windows and a sun visor at the front. The transmission was a specially developed five-speed gearbox allowing on-the-move shifting, the power unit was the Minneapolis-Moline four-cylinder KED series petrol engine which was tuned to run on leaded petrol, and electric starting was included in the standard specification. A belt pulley was provided for stationary work. 'Just as the city man needs a comfortable closed car to pursue his activities, so the farmer who spends a big share of his time on a tractor needs and wants greater comfort on the job,' it was confidently asserted in the UDLX sales leaflet.

There were, inevitably, some disadvantages, and one of these was the placement of the doors at the rear of the cab. Using the back doors did not encourage dignified entries and exits, particularly with equipment such as a wide cultivator hitched to the drawbar. Another problem was the price. The UDLX engine produced about 40HP, and the tractor's list price was £2155; for the same price, farmers could buy two 40HP John Deere or International tractors with a sprung seat and no cab. In spite of an advertising campaign drawing attention to the high specification and the safety equipment, sales were disappointing. Production finished in 1940 after about 150 of the tractors had been sold – a modest total which probably means there were large numbers of disappointed tractor drivers on American farms. In fact, many of the UDLX tractor sales were outside the farming industry, involving customers who used them for industrial or transport work. In financial terms, the Comfortractor was almost certainly an expensive failure for Minneapolis-Moline, but it deserves to be remembered as a bold and imaginative attempt to offer tractor drivers significantly better standards of safety and comfort at least 40 years before the market was ready for such extravagance.

SLOW PROGRESS

The disappointing response to the UDLX tractor, which probably had more to do with the price than with the tractor's concept, did little to encourage other manufacturers to follow the Minneapolis-Moline example. Progress was slow and hesitant, although padded seats did become more widely available from the 1950s onwards, while power steering and more comfortable seats filtered down from the bigger tractors. The acceptance of safety cabs in Europe provided the opportunity for manufacturers to fit heaters and radios, and these became widely available in the 1970s. These were followed by air conditioning, which filtered down from combines and high-horsepower tractors during the 1980s.

It was the 1980s that brought the first significant moves to develop tractor suspension systems. Front-axle suspension has made an appearance from time to time; during the 1950s, several European manufacturers, including the Steyr company in Austria, featured this type of suspension. The sprung axles were abandoned, however, when models fitted with this feature failed to attract enough customers. But the tide was beginning to turn. A large-scale survey carried out in Britain in the 1970s by the National Agricultural Advisory Service (now known as ADAS) and funded by UK tractor manufacturers compared work rates actually achieved by tractors on farms with the work rate of which they were potentially capable. When the manufacturers' representatives drove the tractors at public demonstrations and ploughing matches, the machines regularly achieved their maximum work rate or power utilization. Yet, when the same measurements were taken on commercial farms, power utilization for 65HP tractors averaged 86 per cent, falling to only 55 per cent for tractors of 95HP.

Renault revolutionized driver comfort when it introduced cabs with a full suspension system in the 1980s. The Hydrostable cab also allowed faster working speeds for some tasks, and it is now available as standard equipment or a popular option on many Renault models, including the Ares series.

vibration coming up from the tractor wheels, improving comfort and encouraging faster working speeds. Although the idea attracted some interest, tractor manufacturers were not enthusiastic. A senior executive of one tractor company is reported to have said that he would be unwilling to encourage faster working speeds because the extra vibration on the unsuspended parts of the tractor could lead to reliability problems and the risk of increasing the warranty costs.

Right: Most manufacturers introduced front-axle suspension systems around the turn of the century. Early converts were John Deere tractors equipped with TripleLink suspension (TLS). Its main components are the vertical rams and gas-filled steel accumulators seen on this 6910 model.

Above: JCB's Fastrac tractors have a suspension system for both the front and rear axles. For safety reasons, this is a legal requirement in many countries for high-speed tractors such as the Fastrac, but drivers say it also improves driver comfort and makes their job less tiring.

THE COMFORT–PERFORMANCE LINK

In the 1960s, 95HP was a big tractor, and the fact that drivers were typically achieving little more than half the available work rate was a cause for concern. Customers were not obtaining the performance they had paid for, and it was important to find out why. The drivers were questioned about their driving performance, and the survey found that they were deliberately choosing a lower gear ratio and/or a slower throttle setting, as the tractor became too uncomfortable and noisy if the work rate was increased. The manufacturers' representatives, usually driving for relatively short periods, were prepared to tolerate the discomfort in order to produce an impressive performance. The survey concluded that, in contrast, vibration levels and excessive noise were the main reasons that drivers on farms chose to drive more slowly, and it also concluded that providing additional engine power only made the problem worse.

An obvious deduction was that, in addition to making life more agreeable for the driver, improved tractor comfort would also boost efficiency and profitability. At the same time, there was growing evidence from surveys in several European countries that traditional tractors could be bad for the drivers' health. Too much vibration and badly designed seats caused back problems and internal disorders, and drivers who had been exposed to excessive noise levels during long years at the controls of a noisy tractor were more likely to suffer from hearing disorders.

In the early 1980s, engineers at Britain's Silsoe Research Institute (SRI) developed what was probably the world's first tractor cab with a full suspension system. It was designed to isolate the driver from the worst of the

JOHN DEERE 6910

Manufacturer: John Deere & Co.
Model: 6910
Production started: 1998
Power unit: Six-cylinder turbo, 6.8 litres (408 cu. in) capacity
Power output: 135HP
Transmission: PowrQuad with up to 24 speeds or AutoQuad

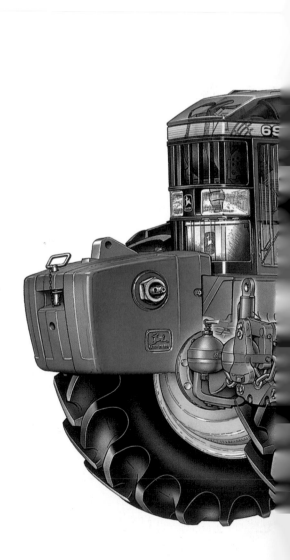

THE HYDROSTABLE CAB

Although the SRI suspended cab failed to reach the production stage, it may have had some influence on a French project that achieved a more positive result. In 1977, Renault Agriculture, the tractor division of the car and commercial vehicle company, had begun work on a cab suspension project; in the same year, engineers from Renault had visited the SRI to exchange ideas on how suspension systems could be designed.

Renault was also developing suspension systems for truck cabs, and the two research teams were able to work together. In spite of this advantage, it took 10 years to complete the development programme. The production version was called the Hydrostable cab, and it was announced at the SIMA machinery show in Paris in 1987 as standard equipment on Renault's high-specification TZ series tractors.

Renault engineers developed a combination of coil springs, anti-roll bars, shock absorbers and transverse rods to produce the suspension system for the Hydrostable cab. It is claimed to have at least some effect on all of the five types of movement tractor drivers are subjected to during their work. These are vertical movements from the tractor tyres, longitudinal movements caused mainly when trailed implements are towed at speed, lateral movements linked mainly to bigger tractors on large tyres, and pitching and rolling movements.

Although the new cab suspension

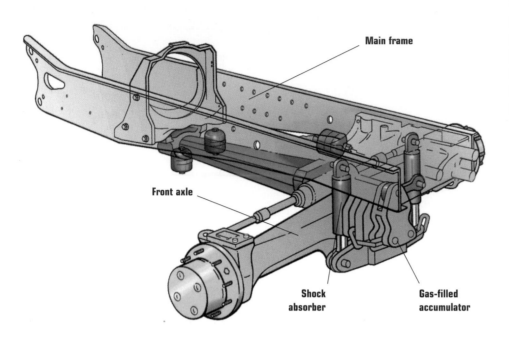

Main frame

Front axle

Shock
absorber

Gas-filled
accumulator

adjustment which enables the driver to
alter its responsiveness, and it added a
front-axle suspension option for some of
its tractors in the year 2001.

FASTRAC SUCCESS

With Renault's Hydrostable cab established, the next important development
in driver comfort arrived in 1991 when
the Fastrac range of tractors arrived
from the British construction equipment
specialist JCB. Production started in 1991
with the Fastrac 120 and 145 Turbo
models powered by Perkins engines. The
JCB engineers chose an unconventional
design for their first venture into the agricultural tractor market, and the list of
features included a gearbox with a top
speed of 72km/h (45mph) for transport
work, four-wheel drive through equal-
diameter wheels, four-wheel braking
and a suspension system over both the
front and rear axles. A four-wheel
steering system has more recently been
introduced for some Fastrac models.

attracted enthusiastic comments from
some users and won a number of
awards for innovation, including a gold
medal from the Royal Agricultural
Society of England, sales remained
disappointing at first. After eight years,
only about 15 per cent of UK customers
were specifying the Hydrostable cab on
their new Renault tractors. Since then,
however, demand has increased sharply;
by 1999, almost 80 per cent of the
Renault Ares models sold in the United
Kingdom were equipped with the
suspended cab. Sales have been helped
by the results of a series of tests carried

out in Germany by the DLG research
organization in 1996. These compared
two 145HP Renault tractors, one with a
Hydrostable cab and the other with a
standard cab conventionally mounted on
rubber isolating blocks. The tests showed
that vibration levels on the tractor fitted
with the Hydrostable cab were reduced
by up to 35 per cent in some situations.
Also, when driving over rough surfaces at
high working speeds, it took up to twice
as long for the driver in the Hydrostable
cab to experience symptoms of stress.
Since then, Renault has improved the
Hydrostable suspension by providing an

With its high top speed on the road,
the Fastrac fits into the exclusive 'fast
tractor' category in which four-wheel
braking and a front and rear suspension
system are legal requirements. The front
suspension is based on coil springs with
twin tube telescopic shock absorbers.
The suspension at the rear is hydro-
pneumatic to give a self-levelling
action, which keeps the chassis at the
same height above the axle regardless of
the weight being carried on the rear
linkage or the load space behind the
cab. The self-levelling action plays an
important part in the Fastrac's ability to

work with draught implements such as mounted ploughs, as well as offering high speeds for transport work. Without the self-levelling effect, the weight of a heavy implement on the rear linkage would compress the suspension and make it difficult to control that implement's working height.

In many countries, an all-round suspension is a legal requirement on 'fast tractors' for safety reasons. The ability to absorb the worst of the shock loads as the tractor wheels drive over a bumpy surface helps the driver to maintain full control of the steering. The Fastrac suspension, like Renault's Hydrostable cab, soon earned a reputation for a smoother ride and high standard of driver comfort.

The extra comfort meant that drivers were willing to adopt faster working speeds where appropriate, as had been originally suggested almost 30 years previously in the original advisory service survey of tractor performance.

New tests carried out by ADAS with Fastrac tractors suggested that the smoother ride meant that working speeds could increase by more than 20 per cent for jobs such as secondary cultivations. The suspension system has undoubtedly made an important contribution to the success of the Fastrac as a fieldwork tractor, in addition to its role as a high-speed transport tractor.

SUSPENSION TAKES OFF

It was inevitable that the success of the Renault and JCB suspension systems would attract the attention of other tractor manufacturers; by the late 1990s, there was a surge of front suspension developments involving most of the leading manufacturers. John Deere's TripleLink Suspension (TLS) arrived in 1997 when its new 6010 series tractors were announced; since then, it has been extended to other models in the John Deere range. TLS uses two double-

The development of rubber tracks has been described as the most important technological breakthrough ever made in crawler tractor technology. The tracks, made of rubber reinforced with steel cables, were first introduced by Caterpillar on its Challenger 65 tractor.

acting hydraulic rams mounted on the main frame of the tractor and linked to two gas-filled accumulators. There is also an axially mounted arm linking the axle to the frame, and a Panhard rod guides the axle laterally. TLS allows 100mm (3.84in) of vertical travel and there is a control system to lock the suspension automatically at speeds below about 1.5km/h(1mph) for safety.

A more recent addition to the list of front suspensions is the Aires system available on some models in the Valtra range of tractors from Finland. The front axle is mounted on a pair of swinging

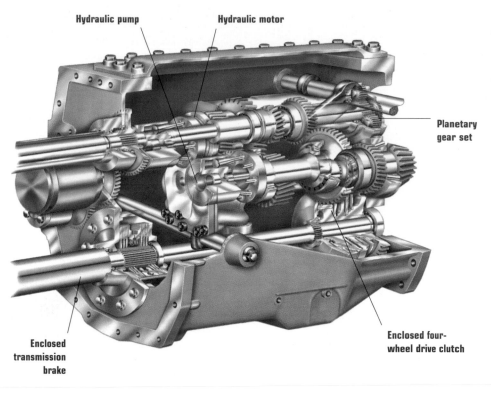

Hydraulic pump   Hydraulic motor

Planetary gear set

Enclosed transmission brake

Enclosed four-wheel drive clutch

A sectioned diagram of the Vario transmission from Fendt, the first of a new generation of German-designed tractor transmissions. It combines both mechanical and hydrostatic drive elements and includes a hi-tech electronic control system.

supply of warm water for hand washing – a convenience factor that deserves to be more widely available.

The non-stop success of four-wheel drive during the second half of the twentieth century was bad news for the crawler tractor salespeople. Farmers and contractors who needed extra pulling efficiency could turn to four-wheel drive tractors instead of trying to cope with the disadvantages of the traditional tracklayer, and sales slumped. Annual production of crawler tractors in the United Kingdom totalled more than 2000 per year in the early 1970s, with more than 50 per cent of them destined for export. The next 20 years brought a rapid downturn, however, and annual sales of tracklayers in the United Kingdom, including imported models, had reached their lowest ebb at fewer than 50 tractors per year by the mid-1990s. If the decline had continued, new crawler tractors would soon have qualified for endangered species status, but, instead of fading out of existence, tracklayers have staged a significant

recovery. By the turn of the century, they were recognized as the star performers in an otherwise gloomy tractor market in Britain and other countries.

REVIVING THE CRAWLER TRACTOR

The development that revived the tracklayer's fortunes was the introduction of rubber tracks to replace the traditional steel version. The idea was developed in the United States by Caterpillar, and the first production tractor equipped with the Mobil-trac drive system with tracks or belts made of rubber reinforced with flexible steel cables was the Challenger 65 tractor. Caterpillar introduced the Challenger 65 in the United States in 1987, and it arrived in Europe in the following year. The 65 was powered by a six-cylinder Caterpillar 3306 series engine producing 285HP and providing 216HP at the drawbar, but it was the Mobil-trac rubber tracks that were the significant feature of the new tractor, and the initial response was sceptical.

Rubber tracks retained all the practical virtues of the traditional tracklayer,

including the unbeatable pulling efficiency in difficult soil conditions. The large contact area of the tracks on the soil surface helped to spread the weight of the tractor and reduce the ground pressure and the risk of soil damage. But the Challenger's new tracks also overcame the problems of slow working speeds and lack of road use that had already persuaded many thousands of farmers to switch from the traditional steel tracks to a four-wheel drive tractor.

Much of the initial scepticism was due to doubts about the durability of rubber

tracks when transmitting large amounts of engine power, and many farmers decided to wait to see if the Mobil-trac system really could stand up to frequent journeys on metalled roads without rapid and expensive track wear. In many cases, the performance of the rubber tracks probably exceeded expectations. They have proved their ability to turn large amounts of engine power into drawbar pull, and the tracks are surprisingly durable when used on the road. The Challenger range now extends to 400HP-plus, and the tractors are

available throughout Europe and in some other overseas markets under the Claas name and colours. The deal with Claas is reciprocal, as it gives Caterpillar the marketing rights to some Claas combine harvesters in North America and some other markets.

The success of the Challengers and their rubber tracks soon brought competition. In Britain, the Track Marshall company based in Gainsborough, Lincolnshire, offered the TM200 tracklayer with an Australian-designed rubber track system. A 200HP

Fendt introduced the Vario transmission on the high-horsepower 926 model, but since then the new drive system has been moving down the power scale. At the time of writing, the 400 series Fendt tractors powered by four-cylinder engines are available in four sizes, with power outputs from 85 to 126HP.

Cummins engine powered it, and the hydraulically operated steering was controlled by a steering wheel instead of the traditional levers. A special design feature was a sideways-tilting cab, operated manually by two hydraulic rams to give access to the area below the cab floor for servicing and repairs.

The TM200 was soon phased out after sales proved to be disappointing, but there were other rubber-tracked rivals that were more successful. They included Case IH's Quadtrac models based on the high-horsepower Case Steiger tractors with articulated steering, but with four rubber track units instead of four driving wheels. The Quadtrac models were uprated in the year 2000, and the improvements list included a power boost to bring the output of the STX 440 model at the top of the range to 440HP.

John Deere moved into the market with rubber-tracked versions of its 8000 series rigid-frame tractors. The range has more recently been extended with the addition of rubber track conversions of the pivot-steer 9000 range with power outputs up to about 425HP. Japan's Morooka range of crawler tractors was among the earliest converts to rubber tracks, and the rubber-track revolution has also spread to other farm equipment, with conversion kits available for machines ranging from

combine harvesters and self-propelled potato harvesters to small two-wheeled pedestrian-controlled tractors.

Conversion kits are also available for other tractors, enabling a well-equipped farm workshop to fit the rubber tracks for ploughing and other heavy draught work when ground conditions are wet and crawler tracks are most beneficial. Rubber tyres can then be switched back again for jobs such as road transport and when the working conditions are more favourable. There is also the prospect that the continuing success of the rubber-track revolution will bring more tractor manufacturers into the market, and AGCO has already shown a prototype version of a Massey Ferguson tractor equipped with a set of rubber tracks.

AIR VERSUS LIQUID COOLING

Engine developments during the period from 1975 included a continuing increase in the use of turbochargers and intercoolers, increasing use of electronic control systems and a virtual end to the competition between liquid- and air-cooling systems. North American

manufacturers have never shown much enthusiasm for air-cooled tractor engines, and liquid cooling has dominated the industry there since the early years of tractor development. Apart from a small number of low horsepower models, British tractor designers sided with their US counterparts and preferred liquid cooling, but the situation in the rest of Europe has been less clear cut. By the late 1970s, Deutz and Same, two of the most influential tractor manufacturers in Europe at that time – now linked as members of the Same Deutz-Fahr tractor and machinery group – were specializing in air-cooled engines for their tractors, while most of their rivals had standardized on liquid cooling.

Liquid cooling does have some disadvantages. The cooling system includes a bulky radiator, pipes, connections, a pump and a cooling fan with a belt drive, and these are all items that can lead to mechanical problems and reduced reliability. According to the Deutz publicity in the early 1980s, 40 per cent of all the failures of water-cooled engines are due to the cooling system. Other problems with a radiator-based cooling system are the risk of expensive frost damage in cold weather and the fact that a liquid-cooled engine also takes slightly longer than an air-cooled unit of similar size to reach its optimum working temperature. A slower warm-up period allows more time for condensation to form on the cylinder walls, and this can lead to problems.

Advocates of liquid cooling, championed particularly by Perkins Engines and Ford, claimed significant benefits for their system. Having a big volume of liquid – normally water with added antifreeze and corrosion inhibitors – in the cylinder block provides a sound-absorbing effect, and even the most loyal fans of air cooling

The engines for the Fendt 400 series tractors are liquid-cooled. This picture shows how the cooling system has been shaped to accommodate the sharply sloping bonnet line that is a distinctive feature of the current Fendt tractor range. The rear-hinged bonnet gives easy access for routine service checks.

Section of the four-cylinder engine of the 400 series Fendt tractors. Turbocharging is standard throughout the range, and the design includes four valves per cylinder for faster gas exchange. The injection system is designed to operate at high pressure to improve the combustion efficiency.

following the trend that was already well established in the European car industry, where millions of Volkswagen Beetles, Europe's most successful car, had demonstrated the effectiveness of air-cooled power.

ENVIRONMENTAL CONSIDERATIONS

Exhaust emission regulations have dominated the development of new tractor engines since the late 1980s, as the designers struggle to keep pace with ever stricter legislative control. The limits are genuinely difficult to meet, and most of the leading engine manufacturers have been forced to scrap existing designs and introduce new power units capable of reaching the required performance levels. The cost of the research and development work and of bringing the new 'clean' engines onto the production line is enormous, and it is a cost that is inevitably passed on to the farmers and contractors who buy the tractors.

Because of the numbers involved, tractors make a relatively small contribution to the global total of engine exhaust emissions and a correspondingly small contribution to air pollution and environmental damage, so the cost of cleaning the emissions is high in relation to the problem it solves. The consolation for customers who have to foot the bill is that cleaner engines are generally more fuel efficient. High levels of carbon and other particles in the exhaust gases are often a result of inefficient or incomplete combustion, and improving the combustion process helps to extract the maximum amount of energy from each tankful of fuel.

Adding a turbocharger, often with an intercooler or charge cooler as well, helps to boost the power output from an engine. It also makes a contribution to cleaning the exhaust, and the emphasis on meeting the emissions legislation has

will sometimes admit that their engines are noisier. Liquid-cooling advocates also claimed a steadier temperature with less extreme fluctuations because of the stabilizing effect of the large volume of cooling liquid, and this can be especially beneficial in situations demanding stop-and-start driving, where the engine temperature alternates between peaks and troughs.

Fuel efficiency figured prominently in the cooling system debate, with claims and counterclaims on both sides. Higher fuel prices had encouraged the tractor engine manufacturers to take fuel efficiency more seriously, although there is little evidence that farmers then or now put it high on their list of tractor priorities. The arguments about fuel efficiency remained inconclusive, but

there were other concerns looming for the advocates of air cooling, or air and oil cooling – as Same insisted on calling it – as the sump oil plays a significant role in removing heat from an air-cooled engine.

One of the developments was the demand for tractors with more power, and cooling by air becomes less efficient as the size of the engine increases above about 150HP. The other factor that gave liquid cooling a significant advantage is the increasingly stringent legislation in place to clean up engine exhaust emissions in Europe and North America. Liquid cooling offers small but significant benefits in this area of engine performance. During the 1990s, the number of tractors equipped with air-cooled engines dwindled steadily,

encouraged tractor manufacturers to include turbochargers as almost standard equipment on engines above about 100HP. They are also widely used on engines in the 60HP to 100HP range.

ELECTRONIC CONTROL SYSTEMS

Another trend in engine design since the early 1980s has been a big increase in the use of electronic control systems on tractor engines, and on some of the biggest engines this can now amount to an almost totally computerized engine management system. Electronics have been available on farm equipment since the late 1970s, when they were used on

complex machines such as combine harvesters to provide the driver with a comprehensive information system on the operation of key components, using sensors to gather the data. With electronics and computers, it is relatively easy to move from simply conveying and displaying information to processing and using the data in a control function. In spite of some misgivings from farmers who were concerned about the reliability of this type of equipment when exposed to the dust and vibration associated with field machinery, electronic information and control systems have provided one of the

most rapidly developing areas of farm mechanization.

The biggest breakthrough in the use of electronics on tractors was the introduction of the Massey Ferguson information and control systems. They were announced in 1986, providing the driver with a comprehensive data display as well as automatic control systems including full electronic control of the rear linkage. The Datatronic II version available in the late 1990s monitors 20 different functions and includes programmable memories, plus data transfer by smart card or through a hard-copy printout facility. An additional function is

automatic wheelslip control, comparing the apparent travel speed, as indicated by measuring the driving wheel rpm, with the actual forward speed measured by a radar unit. This constantly monitors the amount of wheelspin, and the control system can automatically adjust the working depth of the plough or cultivator to reduce the pulling power needed and therefore reduce the wheelslip. Although Massey Ferguson leads the way with the use of electronic controls through the development of its Autotronic and Datatronic systems, other manufacturers have also introduced comprehensive systems producing large

Left: The ultimate in two-way or bi-directional tractor control is provided by the Claas Xerion tractor with its movable cab. The cab can be mid-mounted to face forwards over the engine compartment. It can also be swivelled through 180 degrees and relocated to the tractor's rear to leave a large area for carrying mounted equipment.

amounts of performance-related data and computer control systems.

Massey Ferguson also pioneered another major hi-tech development in agricultural mechanization. This was the use of signals received from the global positioning system (GPS) network of space satellites to identify the position of a machine such as a tractor or a combine harvester as it travels to and fro across a field. The GPS satellite network, more than 19,300 kilometres (12,000 miles) above the surface of the Earth, is maintained by the United States for military purposes. It is also used for a wide range of other functions, and Massey Ferguson began its research programme in the mid-1980s using a signal receiver on a combine harvester.

Above: Moving the Xerion cab is a push-button job. The complete cab unit is raised hydraulically and then turned through 180 degrees and moved to the required position. This hi-tech sophistication adds flexibility for operating many types of equipment.

The combine harvester-based GPS application, announced in 1991, allows combine movements in the field to be tracked accurately within a metre or so, and this data is entered on a map together with information from the harvester's yield-recording equipment. All the data can be transferred by a smart card to the computer in the farm office, where it is displayed as a map of the field showing how the crop yield varies in different parts of the field. This information can be used to identify areas where lower yield may be caused by problems such as compacted soil or poor drainage, allowing remedial action to be taken.

Tractors equipped with GPS receivers and equipment for data processing can be used to apply inputs such as seed and fertilizer, using the yield map information to adjust the application rate to match the yield variations. Field maps can also be prepared showing areas of weed infestation to allow the sprayer to be controlled automatically, applying herbicide on the problem areas but reducing the dose where the field map shows few weeds are present. Using GPS has allowed the development of what has been called 'precision farming', allowing farms to be managed by the square metre instead of on a field scale. As well as the economic advantages this provides, it can also bring environmental benefits through more precisely targeted use of chemical inputs such as pesticide sprays and fertilizers.

THE CVT

Another aspect of the electronic revolution that had a profound impact on the farming industry during the 1990s was the development of a new generation of tractor transmissions. Constantly variable transmissions (CVTs) are a German development, and they offer the first effective challenge to the supremacy of the powershift for mid- to high-horsepower tractors. They also offer some of the benefits of the hydrostatic drive system described in chapter 4, including infinitely variable adjustment of the forward speed without altering the engine rpm.

CVTs were developed by three German companies working at about the same time, and on a basically similar approach to transmitting power from the tractor engine to the wheels, but apparently all working independently. The solution they all came up with is based on a combination of a mechanical drive with gears and the hydraulic pump and motors of a hydrostatic drive system. There is also a hi-tech electronic control system that automatically allocates more of the power to either of the two transmission components to suit the job the tractor is doing. In addition, the electronic control systems for the transmission and the engine are linked, allowing both to respond to changes in the workload. Benefits of the new generation transmissions include: smoother, more flexible power delivery with easy-to-operate controls; higher mechanical efficiency than a hydrostatic drive system which is close to the standards reached by a typical powershift; and an electronic control system that can be more closely integrated with that of the engine.

The first company to offer a CVT on a production tractor was Fendt, now part of the AGCO group. Their Vario transmission was available in 1995, initially on a new 260HP tractor, but since then the sales success has encouraged Fendt to offer versions of the Vario drive system throughout the model range from 86HP upwards. Claas, noted mainly for its harvesting machinery, introduced its HM8 drive system on the 300HP Xerion bi-directional tractor. The HM8 is also a CVT, and it offers eight gear ratios, as well as the stepless hydrostatic drive. Claas is offering the HM8 transmission to other manufacturers for use in construction equipment. The third and potentially the most important of the CVT pioneers is the German-based ZF company. It specializes in making transmissions and other components for the tractor industry, including many of the leading European manufacturers, and the first customer for its new CVT was Steyr in Austria, now linked with Case IH as part of the CNH Global tractor and machinery group. New Steyr and Case tractors with the ZF CVT, covering the 120HP to 170HP output range, were announced in 2000. Many of the other leading tractor

This bare chassis view of a Mercedes-Benz MB-trac systems tractor shows the rugged main frame that carries the cab and includes a coil suspension system. MB-tracs include a number of advanced design features, including a reversible driving seat.

companies, including the Deutz range from Same Deutz-Fahr and the Valtra company in Finland, announced plans to introduce CVT models, using versions of the ZF transmission, and, in 2001, Deere announced it was adopting a ZF-designed CVT for its new mid-range tractor series.

INDUSTRY CHANGES

While much of tractor history deals with technical developments and the way in which new ideas are introduced and adopted, the structural history of the tractor industry is also important. This applies particularly to the changes that have taken place during the past 25 years or so. This has been a period when tractor manufacturers have enjoyed some periods of prosperity, but there has also been intense competition for sales at a time when some important markets were shrinking and leaving the industry with too much manufacturing capacity.

One result of the competitive pressures facing the tractor manufacturers has been a series of takeovers, mergers and bankruptcies that have seen many famous companies either disappear or lose their independence as the industry is consolidated into just a few giant international groups, often controlling numerous brand names. This is a trend that started in the early 1920s when pressure from the then recently introduced Fordson Model F helped to

destroy large numbers of the tractor companies that had mushroomed into existence in the tractor boom during and immediately after World War I.

Some of the tractor companies that disappeared in the 1920s phase of consolidation, particularly in the United States, were small and lacked the resources to provide good after-sales support for what were often badly designed oddball products. There would have been few regrets when they ceased trading or switched to making other products. There was also a rush to move into the tractor industry after World War II, but this time it was on a smaller scale and much of the activity was in Britain. Many of the new British companies survived for just a year or so, and in some cases for just a few months, but even the very successful newcomers such as Ferguson and Nuffield had lost their independence by the end of the 1980s.

The recent history of the CNH Global group, one of the world's top tractor, farm machinery and construction equipment manufacturers, shows just how complex the takeover and merger trail can become. The 'C' in CNH stands for Case, one of the oldest names in the tractor industry. In 1967, the company that then owned Case was taken over by the Texas-based Tenneco group, and, in 1972, Tenneco also acquired David Brown Tractors in the United Kingdom. The David Brown name later

This 100HP Mercedes-Benz MB-trac illustrates the versatility of the systems tractor. It has p-t-o powered equipment mounted on the front and rear linkages, and the load platform behind the cab is carrying a mounted hopper containing additional supplies of seed or fertilizer.

disappeared from the tractor range, but another takeover in 1985 added the International Harvester company to Tenneco's agricultural division. This time the additional brand name survived with the introduction of a new Case IH brand. In 1986, the Steiger Tractor Co. was purchased to give Case IH a vital stake in the important high-horsepower sector, and the leading Austrian tractor maker, Steyr, became the next takeover target in 1998.

'NH' in the CNH name stands for New Holland, another reminder of North America's nineteenth-century farm equipment industry. The long-established New Holland brand name was acquired by Ford in 1986 through a takeover designed to add a comprehensive range of harvesting machinery to Ford's tractor and construction machinery interests. In 1987, Ford made a successful bid for the Canadian-based Versatile company which brought high-horsepower, tractors into the group. The next big development came in 1991 when the Italian-based Fiat company acquired

Ford's New Holland agricultural and construction equipment interests and merged them with its own Fiatagri business to form New Holland Geotech. CNH Global was formed in 1999 by merging the Case IH and New Holland operations, with Fiat retaining a controlling financial interest.

CNH Global is one of the big four companies that currently dominate much of the tractor industry worldwide. The other members of this exclusive big league are: Deere & Co., the US manufacturer of John Deere products; AGCO, building tractors under the AGCO Allis, Fendt, Massey Ferguson and White brand names; and the Italian-controlled Same Deutz-Fahr group, which builds agricultural tractors in Italy and Germany, as well as a range of compact tractors in Poland. By the turn of the century, the big four companies were dominating the tractor market. In Britain, for example, they accounted for four out every five new tractors sold in the 40HP-plus sector, which covers most of the tractors sold for farm work, but does not include the compact models sold mainly to golf clubs, local authorities and other amenity users.

The stream of takeovers and mergers means there are only a small number of long-established independent companies still operating on a substantial scale in North America and Europe. The European list would include Valtra (previously Valmet) in Finland, Landini in Italy and Renault Agriculture, part of the French government-owned Renault organization. There are also some smaller specialist companies such as Lindner in Austria and the Italian Carraro company, and some of the tractor companies in Eastern Europe, including Ursus and Zetor, produce tractors in large numbers.

Tough competition and the large number of companies that have been taken over or have ceased trading can make tractor production seem an unattractive business. Despite the problems, some companies have recently moved into the industry for the first time. An interesting new arrival in Germany is the Doppstadt tractor range, designed as updated versions of

McCORMICK MC115

Manufacturer: McCormick Tractors International
Model: MC115
Production started: 2001
Power unit: Four-cylinder turbocharged Perkins engine
Power output: 115HP
Transmission: Powershift/ powershuttle with 16 forward speeds (32 with the creeper gear option)

the MB-trac, a systems tractor developed and built by Mercedes-Benz. The MB-trac was highly successful until Daimler-Benz decided to pull out of the farm tractor market in 1991, and the Doppstadt versions have inherited some of the MB-trac features. The Trac 160 model is powered by a 160HP engine and features a shuttle transmission with 40 speeds forwards and in reverse, and four-wheel steering is on the options list.

NEW BRITISH ARRIVALS

Britain, the world's biggest tractor exporter during much of the twentieth century, has produced several new arrivals since 1990, as well as a 'new' brand name. The brand name is, in fact, one of the oldest in the industry. The name chosen for the new tractors being built at the former Case IH factory in Doncaster, Yorkshire, is McCormick, after one of the two old-established American machinery firms that joined forces in 1902 to form International Harvester. The name started with Cyrus H. McCormick, who designed an improved reaper and demonstrated it in 1831, taking out a patent in 1834. The success of the reaper helped McCormick to build up a highly successful machinery company, and, after the merger with Deering, his name continued to appear on many of the tractors built by International Harvester.

It was International Harvester that had opened the tractor plant in Doncaster, and the first tractor to roll off the production line there in 1949 was a

McCormick Farmall Model M. Case acquired the plant when International Harvester was taken over, and the compulsory sale of the factory was one of the anti-trust conditions imposed when the Case/New Holland merger forming CNH Global was approved. When the sale was completed in 2000, the new Italian owners, who also own the Landini tractor business, chose the McCormick brand name for the new Doncaster-built tractor range.

Another anti-trust condition imposed when the merger that formed CNH Global was approved was the compulsory sale of New Holland's Versatile plant in Winnipeg, Manitoba. Case already owned the Steiger factory,

and retaining Versatile would have left CNH with a dominant position in the high-horsepower sector. The company that bought the Winnipeg factory was Buhler Versatile, and it is continuing to make tractors for CNH, as well as developing its own brand.

Two of the British new arrivals are unconventional tractors with distinctive design features which have helped to establish them. The Fastrac high-speed tractor range from JCB Landpower, available since 1991, has already been referred to because of its innovative suspension system. The Multidrive from Thirsk, Yorkshire, also has a front and rear suspension system, but in this case based on coil springs at both ends.

The company that owns the Landini tractor factory revived the McCormick brand name when it bought the Doncaster, Yorkshire, tractor factory from Case IH. Cyrus H. McCormick was one of the greatest names in the history of American farm machinery development.

When the first version of the Multidrive arrived in 1992, the year after the Fastrac launch, it was called the Clayton Buggi and was designed and built by Lucassen Young of Stockton-on-Tees, Cleveland. It was based on a chassis with four-wheel drive through equal-sized wheels, four-wheel hydrostatic steering and a large load area behind the cab. An automatic trailer braking system was provided, but the three-point linkage at the rear was an optional extra.

Lucassen Young designed the Buggi mainly for carrying a demountable sprayer, but the load space can also be used for other jobs, including carrying seed for a rear-mounted seed drill. It is not, however, designed as a heavy-duty pulling tractor for jobs such as ploughing. A 110HP John Deere engine powered the original version, but the range was later extended to include more powerful engine options and a version with an extended wheelbase was introduced to allow more space on the load platform. The name Buggi was later changed to C-trac. In 1999, the business was taken over by Multidrive of Thirsk,

THE FUTURE OF POWER FARMING

Predicting future developments in the tractor industry has never been easy, and ideas forecast by tractor designers in the past have often failed to materialize. Further improvements in driver comfort during the next decade are likely, and the rapid progress in electronics and computer control systems means the fully automated, driverless tractor could be just a few years away. Further on, as the world's oil wells run dry, farmers could grow their own energy crops to fuel their own tractors.

The car industry has a long tradition of producing futuristic models on a one-off basis, the so-called concept cars that draw the crowds at motor shows. They are often simply styling exercises aimed chiefly at attracting publicity, but they can also offer an opportunity to test public reaction to new ideas that may eventually be included on production models. Concept tractors have appeared less frequently, probably reflecting the smaller budgets available in the tractor industry. Some of them were apparently designed with more emphasis on short-term publicity objectives than on long-term market research aims.

One example is the Typhoon II tractor built by Ford in 1965 as a half-size scale model. Like other concept tractors

The vast farms of North America have created a demand for ever larger, more powerful tractors. Manufacturers such as Ford and Renault have been experimenting with ways of making the tractor, and ultimately the farmer, more productive.

Robot control systems for tractors have already reached the prototype stage, and in 2000 Renault provided a public demonstration of its Tractosat automatic guidance system in action on an Ares tractor. It could be available commercially in 2002, but not as a complete replacement for the driver.

improved operating efficiency. Agro Nav uses GPS to track the position of the tractor, and a computer programme automates all the control functions. This allows the tractor to operate without a driver for a wide range of field operations, and the result is said to be higher levels of precision than would normally be achieved with manual control.

The year 2003 is the target date for Renault to introduce its new Tractosat automatic guidance system on a commercial basis. Tractosat, demonstrated for the first time in 2000, shows that Renault has moved away from the voice control system featured in its previous Centaure project. The basis for its new control technology is GPS, which will provide automatic guidance, working either on its own or in conjunction with a computer programme capable of automatically sequencing a series of field operations.

Although Deutz, Renault and the Japanese research centre were the first organizations to demonstrate the new generation of tractor control systems, most of the other leading tractor manufacturers are involved in similar research projects. One example is Deere & Co., which has released brief details of the experimental tractor it is using to carry out research into GPS-based guidance systems. The tractor is a narrow or orchard version of the John Deere 5010 model, and by 2001 it was already capable of operating without a driver.

Any of these programmes could have the potential to produce a driverless tractor on a commercial basis, and this could happen very soon, but the reality is likely to be quite different. A full-scale unmanned tractor would have a number of barriers to overcome, and one of these would be a credibility gap that would see many farmers reluctant to hand over crucially important tasks such as seed drilling, spraying and fertilizer spreading to a robot.

A robot tractor would also pose a number of practical problems. For the foreseeable future, the tractor would almost certainly need human help for the drive to and from the field, and this

suggests that a second person will be needed to ferry the driver back from the field after setting up the tractor for unmanned operation. The driver would also need transport back to the field later to collect the tractor and return it to the farm buildings. There is also a big question mark over who will be available to refill the seed or fertilizer hoppers, or top up the sprayer tank if the tractor does not have a driver, as these tasks would be difficult to automate. Driverless tractors would also pose some difficult safety issues, and there are also legal questions to be resolved before they can be allowed to roam the countryside.

This is why Deutz-Fahr and Renault have both stressed that their new automatic control systems are not designed to replace the tractor driver, but will simply make the driver's job less tiring. According to Renault, 80 per cent of the energy used when driving a tractor is required to operate the steering and the other controls. Handing these tasks over to an automatic control system would allow the driver more time to concentrate on setting and checking the equipment the tractor is powering. This seems an awkward compromise, as it might make the tractor driver's job easier, but it would also devalue it, and many drivers would not welcome a control system that reduces their responsibilities to keeping an eye on the machine at the back of the tractor.

The team behind the Japanese driverless tractor project take a different line. In Japan, there is a serious shortage of tractor drivers, mainly because many young people are not attracted by the career prospects in the farming industry, and a driverless tractor in this country would be specifically designed to replace the drivers who are no longer available. In the long term, this is a more logical way to use the technology available, and sooner or later we will have to come to terms with tractors that do not need a driver or a machine-minder.

FINDING A NEW ENERGY SOURCE

Questions about robotics are likely to be at the forefront of the debate about tractor design and operation for many years, and another increasingly important subject will be the type of fuel we use in our tractors. Finding a replacement for diesel fuel will become a major issue because, at some time during this century, the world's oil reserves are likely to be exhausted. The search for alternative fuels will become increasingly urgent, as indicated in Renault's Centaure project, which suggested solar energy would become the power source for tractor engines

during the twenty-first century.

One of the problems with solar energy is that the amount of power available depends on the hours of sunshine. Presumably the introduction of solar-powered tractors would have to wait until more efficient batteries are available to store the electrical energy produced while the sun is shining, as having to delay an urgent ploughing or seed drilling job until the clouds clear could create problems.

The non-fossil fuel farmers would most welcome is bio-diesel produced from crops such as oilseed rape. Bio-diesel offers significant benefits, as it can be used in most modern diesel engines

Aerials and cameras mounted on the cab roof collect data and images for the automatic guidance system on this New Holland NH2550 self-propelled swather. The experimental machine can guide itself backwards and forwards as it cuts the crop, but needs help finding its way to and from the field.

without modification, and the technology for growing and harvesting the fuel crops is already familiar. It is also environmentally friendly, as exhaust fumes from bio-diesel are much cleaner than those from engines burning petrol or diesel, and producing fuel from annually renewable crop plants reduces the impact on global warming.

Bio-diesel is already commercially available in some countries, but at this stage it is more expensive than diesel fuel, and it has to carry a lower rate of tax to enable it to compete. The economic picture may change in the future as oil prices rise in response to more of the reserves being used, and there is a real possibility that farm-produced bio-diesel could make a small but important contribution to the world's energy needs.

The JCB Fastrac tractor suggests one way in which the tractor industry could develop during the next five to ten years. To achieve a unique combination of fast transport speeds with good performance as a field work tractor, JCB developed a suspension system with a unique self-levelling action.

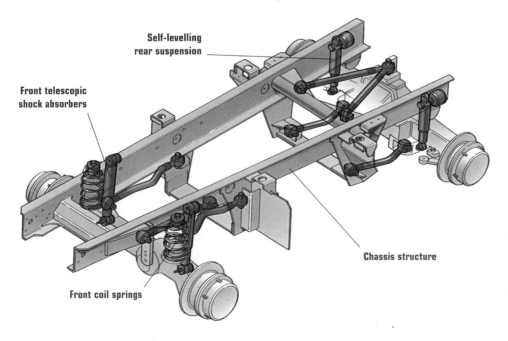

Self-levelling rear suspension

Front telescopic shock absorbers

Chassis structure

Front coil springs

INDEX

Page references in italics refer to illustrations.

PICTURE CREDITS

Amber Books Ltd: 13, 15 (both), 18-19, 20, 21, 22-23 (both), 24, 25, 26, 30-31, 33 (both), 34(b), 35, 36-37 (both), 40, 41, 44, 44-45, 46, 49, 50-51, 56, 58, 61 (both), 76-77, 91, 114, 115, 118-119, 128, 129. **Corbis:** 166-167. **John Deere:** 134. **Mark Franklin:** 44-45, 62-63, 68-69, 82-83, 136-137. **Kevin Jones Associates:** 34, 36, 50-51, 76-77, 132, 135, 138-139, 140-141, 152-153, 158-159, 171. **Andrew Morland:** 48, 78-79, 80, 81, 83, 84-85, 85(t), 87(b), 88-89, 110-11 (both), 112, 113. **TRH Pictures:** 6-7 (Beech). **Michael Williams:** 8, 9, 10, 11 (both), 12, 14, 16, 17, 27, 28, 28-29, 32, 34(t), 38, 39, 42-43, 47, 52, 52-53, 54-55, 57, 59, 60(t&m), 64 (both), 65, 66, 66-67, 70, 70-71, 72, 73, 74, 74-75, 76, 84, 86, 87(t), 92, 93, 94, 95, 96-97, 98-99 (both), 101, 102, 103, 104, 105, 106, 107, 108, 109, 116, 117, 120, 122-123 (all), 124-125 (all), 126, 127, 130-131, 133, 141, 142-143 (JCB), 144, 145, 146 (JCB), 147 (John Deere), 148(t) (John Deere), 148(b) (Case), 149 (Caterpillar), 150-151(m) (Caterpillar), 151(t) (Claas), 152 (John Deere), 154-155 (both) (Fendt), 156-157 (both) (Fendt), 159 (Claas), 160-161 (both) (Mercedez-Benz), 162-163 (McCormick), 164 (Multidrive), 165 (TYM), 168 (Ford), 169 (Kubota), 170, 172, 173 (t).

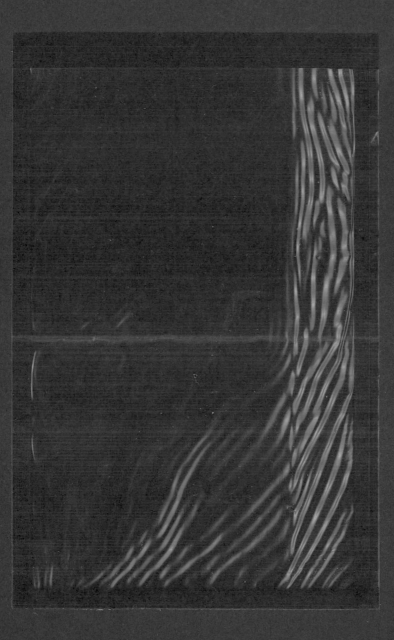